Eloquent Images

Eloquent Images

Word and Image in the Age of New Media

edited by

Mary E. Hocks and Michelle R. Kendrick

The MIT Press Cambridge, Massachusetts London, England

This book was set in Rotis Sans and Janson by Graphic Composition, Inc.
Printed and bound in the United States of America.

Library of Congress Cataloging-in-Publication Data

Eloquent images : word and image in the age of new media / edited by Mary E. Hocks and Michelle R. Kendrick.
p. cm.
Includes bibliographical references and index.
ISBN 0-262-08317-5 (hc: alk. paper)
1. Visual communication. 2. Digital media. 3. Criticism. I. Hocks, Mary E. II. Kendrick, Michelle R.

P93.5.E56 2003
302.23—dc21

2002035062

10 9 8 7 6 5 4 3 2 1

Contents

Acknowledgments

Our deepest thanks go to our contributors, whose excellence made this book possible, to our editors, Douglas Sery and Deborah Cantor-Adams, and to the staff and reviewers at the MIT Press. We also sincerely thank our colleagues who read or discussed the manuscript with us over several years' time—especially Jeffrey T. Grabill and Anne Frances Wysocki. Jenny Wing and Liz Tasker, our research assistants, provided invaluable and cheerful help with reading and preparing the manuscript and the index. A final thank-you to our departments at Georgia State University and the University of Washington at Vancouver for actively supporting our research for this project.

To Richard and Elaine Hocks, for their inpiration.

—Mary E. Hocks

For Griffin Gates and Sofia Grace, my little ones who are much loved.

—Michelle Kendrick

Eloquent Images

INTRODUCTION: ELOQUENT IMAGES

Mary E. Hocks and Michelle R. Kendrick

The ubiquity of electronic media has brought both the enthusiasts and the Luddites out of hiding and sparked debates about how such technologies will affect the scholarship of writing and research methods in the humanities. The emergence of interactive digital media—the "new media" of World Wide Web documents, CD and DVD titles and immersive virtual reality—has some scholars concerned about the encroachment of the visual into territory formerly held almost exclusively by text and print. Some celebrate the growing use of images online, whether photographic, digitally created, animated or visual elements of design, as it suggests our return to a pictorial age, in which knowledge is communicated as often through images as through words (see especially Ong 1982; Bolter 1991; Lanham 1993; and McCorduck 1995). Others worry. These more skeptical scholars suspect images have an inherent conflict with comprehending texts or even possess seductive dangers sometimes cast as feminine. These dangers, some argue, must be controlled and managed, denigrated or downplayed (see especially Postman 1992 and Birkerts 1995). This book takes up the current status of the "eloquent image" by examining rhetorical and cultural uses of word and image, both historically and currently. *Eloquent Images* demonstrates that to attempt to characterize new media as a new battleground between word and image is to misunderstand radically the dynamic interplay that *already exists* and has *always existed* between visual and verbal texts and to overlook insights concerning that interplay that new media theories and practices can foster.

The relationships among word and image, verbal texts and visual texts, "visual culture" and "print culture" are interpenetrating, dialogic relationships. The contradictions, overlaps, and paradoxes inherent in the rhetorical use and interpretation of words and images have been with us since the earliest verbal and visual communication systems; these complex relationships exist in ancient rhetoric and persist in rereadings of the classical rhetoricians, in cultural studies of technology, and even in the binary code distinctions of digital environments. This collection of essays blends theory, critique, and practice to present the assumptions of an interdisciplinary array of artists, critics, and designers about the complex and often contradictory relations of word and image. If the divergent essays by these authors share something specific, it is a call for new approaches to hypermedia design. In issuing such a call, this book responds to current questions as to whether digital media should be understood simply as a pastiche of existing forms of inquiry and communication or whether, in fact, digital media have brought forth a radical paradigm shift that requires new methods of inquiry and understanding. The responses offered by the contributors to this volume show rich and varied possibilities for interpreting and challenging these binary assumptions (new media is not radically different; new media is revolutionary). Hence, the responses argue that there can be no simple, singular approach to new media texts. By bringing cultural, rhetorical, and applied approaches together into one volume, we also help correct the tendency to reduce descriptions of hypertext and hypermedia to a purely formal kind of poetics and aesthetics (e.g., Espen Aarseth's [1997] description in *Cybertext*), to one kind of theory (i.e., the literary theories emphasized by George Landow [1992]), or to any theory of communication isolated from production and rhetorical contexts.

We take up these crucial questions about new media so that we can examine what is at stake for us as artists, critics, designers, and teachers. We believe that understanding the complex history and potential of communication technologies, typically explored in the scholarly literature of science and technology studies, film and media studies, and computers and composition studies, is crucial for *all* teachers and scholars. For better or worse, such technologies are redefining our basic assumptions about reading, writing, communication, and education. It is therefore important to ask: How "new" exactly are the new media? Do these media change our thinking and our work? Scholarly writing has, for some time, debated the "revolutionary" quality of new media, with some theorists embracing and others critiquing the revolutionary assumptions about hypertext and new media as ushering in the "late age of print" or even breaking completely with precedents in print and other older forms of media.[1] Some of the historical continuities between newer and older media have been suggested in Jay David Bolter and Richard Grusin's (1999) influential *Remediation: Understanding New Media.* At the same time, an important online scholarly practice of innovative hypermedia design has been sustained by pioneering hypertext/hypermedia theorists like Stuart Moulthrop, Michael Joyce and Nancy Kaplan, all of whom use electronic media to complicate definitions of literacy, of words and images, and of design. Like these scholars, contributors to this volume complicate the word/image binary through embodied new media practices. We argue collectively in this volume that such practices give us the opportunity to (re)recognize and to theorize the complex relationship that has always existed between word and image.

From Word/Image Binaries to the Recognition of Hybrids

The impulse to see new media and visual culture as radical paradigm shifts from older media and print culture has a clear explanation when we look at the sociology of science and technology. The persistent distinctions historically between "visual culture" and "print culture" are symptomatic of what sociologist Bruno Latour calls modernist thinking: the binary-based thinking that posits radical paradigm shifts from one communications medium to another or from one form of writing technology to another. Latour thus gives us an insight into the history behind the persistent narrative of the binary, explored here as word/image. He describes the "modern constitution" in which "hybrids" are practices that embody complexity and entangle diverse elements. Complexity, in this fashion, can induce widespread cultural anxieties over borders, purity, chaos, miscegenation, and contamination. Latour explains that to assuage such anxieties, we create disciplinary and other dichotomies to purify and avoid the actual hybrid nature of our world and, in this case, the hybrid forms of all media. If we follow this premise, we see historically a tendency to posit radical paradigm shifts as a way to escape cultural anxieties about hybridity, including the impulses to dichotomize our experiences with visual and verbal communication systems.

One of the defining features of the "modern," according to Latour (1993), is that it designates "two sets of entirely different practices which must remain distinct if they are to remain effective" (10). One modern set of practices is the creation of hybrids, the complex mergers of the natural and the human. The alternate set of practices, which Latour calls the work of purification, "creates two distinct ontological zones" (10–11). It is narratives of such purification that establish binaries, which then deny the hybridity that the first set of practices creates. Latour defines this elaborate system of empowerment and denial as the modernist "constitution," whose contradictory work is to enable a proliferation of the very hybrids that it seeks to deny:

> To undertake hybridization, it is always necessary to believe that it has no serious consequences for the constitutional order. There are two ways of taking this precaution. The first consists in thoroughly thinking through the close connection between the social and the natural order so no dangerous hybrids will be introduced. The second one consists in bracketing off entirely the work of hybridization on the one hand and the dual social and natural order on the other. (12)

More often than not, the latter system of denial is practiced. Latour's theory of dichotomies explains the self-referential quality and circular logic seen in much theorizing of visual literacy: Images are "natural" representations; images are created for or by manipulation; images have a grammar, like alphabetic language and yet are not at all like alphabetic language. Such dichotomies, Latour argues, produce compensatory narratives, which enable the modernist constitution to continue to produce hybrids while denying their existence.

The binaries that have been created to describe the new media—linear/hypermediated, visual/textual, image/word, emotional/rational, natural/constructed—are, in Latour's terms, purification narratives. That is, the true hybridity of new media, and all older media, with their interwoven and contradictory mode of being, is pushed aside in such ordering systems. The hybridity of this nascent mode of writing/viewing/experiencing is denied as many theorists and practitioners concentrate on placing their work in one of the two camps. Scholars and designers might ignore the visual elements and discuss the writerly elements, or they might highlight visual elements and either bemoan or celebrate them. As a necessary consequence of operating under the assumptions of a word/image binary, those purification narratives simply obscure the reality and complexity. We can thus trace the hybridity of all written and visual communication systems back to classical rhetoric, and it persists into the age of new media.

When we say that new media are a hybrid, then, we don't suggest any kind of clear resolution of the debates and contradictions inherent in previous media or a comfortable Hegalian synthesis in new media environments. Instead, we suggest a way of parsing components of various media theories and practices as hybrids that, taken as a whole, would make us nonmodern. For Latour (1993), being nonmodern means recognizing these networks of meaning as far vaster and more complicated than the digital forms themselves: "Seen as networks, however, the modern world, like revolutions, permits scarcely anything more than small extensions of practices. . . . When we see them as networks, Western innovations remain recognizable and important, but they no longer suffice as the stuff of saga, a vast saga of radical rupture" (48). Theories of hypertext and new media have sometimes drawn upon this image of radical rupture to historicize what is "new" about new media. The idea of new media as revolutionary, as "radical rupture," has, we argue, been overstated, and chapters in this book follow scholars who favor more cautious, historicized, and situated perspectives.[2] Borrowing Latour's model, we can now define new media as yet another hybrid of word and images, something knowable only in specific local practices and contingent change. In this theory, specific instances become the embodiment of the technology in the

———

moments of design, of rhetorical engagements between actual moments of production and consumption.

As the word/image binary has continued to operate in recent theories of the visual, literature about visual communication has tended to fall on one side or the other of the word/image split, either reducing images to a verbal grammar in which one can become literate (e.g., Donis Dondis's [1973] concept in *A Primer of Visual Literacy*) or idealizing images as unique, holistic truths (e.g., the "pictoral turn" in W. J. T. Mitchell's [1995] *Picture Theory*). The scholarship of composition studies and technical communication has begun to correct these dichotomizing tendencies by demonstrating how visual rhetoric is intertwined with how we construct and analyze texts for particular readers at particular points in history; this scholarship often complicates or even undercuts the abstract and polarized theories of words and images that we have inherited.[3] Along with grounded readings of visual rhetoric, scholars in technology and cultural studies (e.g., Mirzoeff [1998]) add the awareness of all material culture's visual semiotic potential, an awareness that provides a starting point for most chapters in this volume. In the specific readings in these chapters, then, we can see complex, interpenetrating relationships between words and images. Chapters in the opening section of this book introduce and rigorously examine new media theories, artifacts, and design processes as complex hybrid forms and locally situated practices that depend fully on reading and interpretation. The sections that follow explore the complex historical and current uses of images in different communication systems, setting the stage for final chapters on hybrid identities and cultures designed in new media.

Verbal and Visual Practices in New Media

This volume's opening section situates us historically, theoretically, and rhetorically within the most salient scholarship about hypertext and new media and offers three complexly situated approaches to new media theory, critique, and production. In chapter 1, Jay Bolter argues powerfully for the importance of new media design practices in the cultural studies curriculum. Building on his widely influential theories of visual writing spaces and the remediation of older media in new media, he cites the kinds of embodied projects created by faculty and students at the Georgia Institute of Technology and ultimately advocates a kind of embodied cultural theory, one that abandons its claims to being above or outside of practice. Outlining broadly the historical and cultural tensions between print culture and visual culture, Bolter concludes that, with the advent of new media, the ratios have changed to privilege practice over theory, production over critique, formal over ideological, and visual over verbal. He ultimately sets up these dichotomies as heuristics to be subjects for debate as we move into a more complex understanding of new media.

In chapter 2, Anne Wysocki offers thoughtful critical perspectives on the central debates about the effects of visual and interactive digital texts on readers. Wysocki argues against two assumptions in new media studies: that hypertext creates politically engaged and empowered readers and that images weaken readers by making interpretation too easy. By analyzing two pieces of computer-based interactive multimedia, she demonstrates that these new media documents can completely undercut the choices readers feel they can make and that the visual elements of these media can challenge readers in clever ways. Wysocki's readings prompt us to think critically about how power is or is not afforded to readers and audiences, as well as how images and words together complicate and complexify both the message and the experience of new media. Wysocki concludes that the artifacts she analyzes cannot represent all multimedia but instead illustrate the kinds of tensions possible when one is designing online arguments, concluding that "we can compose so differently only if we acknowledge that the visual and hypertextual aspects of our texts are not monolithic."

The final chapter in this section, chapter 3, by Helen Burgess, Jeanne Hamming, and Robert Markley analyzes the production and design of *Red Planet*, a DVD (the primary authors of which are Harrison Higgs, Michelle Kendrick, Markley, and Burgess) that documents cultural and historical approaches to the planet Mars. Burgess, Hamming, and Markley's description and enactment of dialogic cultural analysis concurs with Wysocki's emphasis on the rhetorically sophisticated nature of dialogic texts even as it fulfills Bolter's hopes for an embodied cultural studies. These authors highlight how *Red Planet* functions as a fully integrated form of cultural theory—production and critique—and new media theory in practice. *Red Planet* exemplifies how digital interactive media remediate previous media derived from films, narrative, and scholarly forms. The authors' description of design processes urges them to conclude that "'information architecture' and 'theory' are never distinct concerns . . . . *Red Planet* may serve as a case study in the ways in which 'text' and 'visual images' interact dialogically with the changing technologies (sound, video, and dynamic animation) that are always in the process of redefining the conceptual frameworks and practices of multimedia." A complex and instructive example of new media theory and design in practice, *Red Planet* challenged the designers to find visual strategies that would supplement or enhance the narrative rather than disrupt it and to theorize the economics of video-rich multimedia production. The chapters in this section all ultimately reject persistent and seductive dichotomies, resting instead on complex, uneasy, dialogic understandings of how we interpret and create meaning in new media scholarship.

———

Historical Relationships between Word and Image

As Bolter (1991) documented in *Writing Space*, all words began as images first, and their pictorial quality became more transparent over time. In the volume's second section, the ancient practices of hieroglyphics and classical rhetoric provide evidence of both the continuum and the evolution of textual and visual language practices, from these ancient visual rhetorical systems to present-day digital image representations. By looking at one of the earliest systems of hieroglyphic writing, Carol Lipson shows in chapter 4 how the pictorial systems of ancient Egyptian public texts create multiple "readings"—pictorial, iconic, semantic—that are at once valid and simultaneous. By reading closely the narratives and epideictic messages of these ancient texts, Lipson demonstrates that, although most of these images confirm the values of the elite tradition of the culture that produced them, the system could in fact accommodate multiple points of view and major deviations. She reveals the visual rhetorical system exemplified by the texts as hypermediated by distancing the viewer as a witness to the scene, repurposing the predynastic art of Egypt, and ultimately offering "a rhetoric of accommodation to the ideal [that] could encompass contradictory elements." The multiplicity of available readings, she concludes, is a hallmark of new digital texts that finds precedents in the rhetorical functions and visual elements of ancient writing systems.

Kevin LaGrandeur uses classical rhetoric in chapter 5 to set up images and text as separate means of persuasion that support one other rhetorically, drawing precedents for this activity from classical texts and applying them to Web site design. He points out that "[o]ne [rhetorical] tradition, stemming from Aristotle and continuing with the early Greek orator Gorgias, concerns the affective similarity of images and words . . . . The other tradition, most famously associated with the Roman writer Horace, emphasizes how the poetic image can be persuasive." The principles of using visual displays extracted from classical rhetoric can thus be used to analyze the new media image within the context of a live audience and recapture the image's persuasive power. Because the Web is intertextual, images work via parataxis as a coordinate, supportive structure to textual information, which LaGrandeur demonstrates using sample Web sites and teaching experiences. Understanding the image, according to LaGrandeur, also means comprehending its dichotomous possibilities: Its persuasive power might add to an argument by using ethical, emotional, and logical appeals, but its force and nonrational nature might also distract from a message's logical appeal. Echoing Aristotle's skepticism, LaGrandeur cites examples of unwarranted ascriptions of credibility and dangerous emotional force in a hate group's site, concluding that "[g]raphics sometimes lend undue credibility to otherwise weak arguments."

Taking up the current debates between word and image, chapter 6, by Matthew Kirschenbaum complicates the idea that words and images overlap as Kirschenbaum highlights the clash between critical discourses about images as cultural artifacts and the digital computational processes of electronic data. Working within the field of applied humanities computing, in which scholars build multimedia archives of visual and verbal texts, such as the William Blake Archive, Kirschenbaum demonstrates that "the material truths of digital reproduction exist in constant tension with the Web's siren song of the visual"—material truths that are also well documented in the history of printing—and concludes that "one cannot talk about words as images and images as words without taking into account the technologies of representation upon which both forms depend." As he reads words and images on the micro level, Kirschenbaum complicates our ideas about both the function and the consumption of images, because the deep computational structure of words versus images reveals them as very different on the binary level. Because of underlying mathematical variations and the way that the computer processes such differences, working with images—creating, searching, or manipulating them—differs dramatically from working with text, and these basic differences affect what composers and researchers can accomplish with words versus what they can accomplish with images. Kirschenbaum ultimately suggests that his cases illustrate the continuities between analog and digital reproduction, both of which keep words and images distinct. The chapters in the second edition, taken as a group, assert that in the context of their rhetorical uses within classical and modern persuasion, words and images are not distinct communicative elements; yet paradoxically, they remain distinct in digital media on a fundamental ontological level.

Perception and Knowledge in Verbal and Visual Texts

As each chapter in the book's third section examines the evolving relationships among images, visual culture, and the text-based traditions of research, it provides a cautionary tale about the tensions between visual and verbal textual systems from the distinct theoretical perspectives of cognitive science and cultural studies. Theories of communication have tended to emphasize the textual bases of how we process and understand information, so those who explore the impact of images on understanding reveal some contradictory relationships among the perceived effects of images on readers. These troubled relationships, furthermore, cannot be imputed exclusively to new media but instead take us to the heart of questions about how we make meaning. Each of the critiques presented in this section's chapters suggests how differently we process visual and textual information, and each comes to quite different conclusions about how insights regarding these differences need then to be applied. Thus, Nancy Barta-Smith and Danette DiMarco focus in chapter 7 on how the vastly overstated communications revolutions from print to digital, oral, and visual forms ignore cognitive evidence about communicative acts. Chapter 8, by Jan Baetens, provides a very lucid overview of necessarily contradictory cultural logics of the image when used in modern print texts. In chapter 9, Jennifer Wiley confirms similar results in her applied study of the consumption of image-rich education texts, which causes her to warn teachers about the appropriate contexts for using images and the limitations of "visually rich presentations" in conveying meaning.

Questioning the idea of a visual revolution as radical rupture, Barta-Smith and DiMarco use the developmental cognitive theories of Maurice Merleau-Ponty as evidence that perceptual, visual knowledge has always been primary to both thought and expression. Recent critiques of comparative biology and of Jean Piaget, among others, show that continuity in the field was abandoned for a theory of "watershed moments" of radical change. Readings in biology and cognitive development reveal a naturalized evolutionary metaphor that is more about continuity than change. Similarly, when the authors look at reception of word and image, they see more continuity than change. The authors locate such continuity among oral, print, and visual communication systems in perception. Using insights from other disciplines allow Barta-Smith and DiMarco to critique the tendency of influential thinkers like Walter Ong and Marshall McLuhan to locate communication revolutions in movements from orality to print and to "limit thought to word, oral or print." The authors offer evolving forms of perception as the basis for all communication: "Merleau-Ponty, reminding us of the embodiment we have so often neglected in Western philosophy, tells us that meaning already inhabits things."

Baetens turns to communications and cultural studies as he analyzes the collaboration of media philosopher McLuhan and graphic designer Quentin Fiore on *The Medium Is the Massage*. He sees the collaboration of the two as an important predecessor to postmodern writing that resists textually based logic, and he uses the example of Marie-Françoise Plissart's photographic novel *Droit de regards* to demonstrate how one can substitute a visual discourse for a textural discourse. Since, for Baetens, debates about electronic media are part of a larger debate about visual culture, images that mean to suggest readability, economy of information, or modernity can in practice can lead to a collapse of the meaning of the text they accompany. He offers two functions of images—as pictures of the text, and as representations in themselves—and cites the

McLuhan-Fiore collaboration as the best example in which the image resists being subordinated to the text's meaning, having instead a logic separate from that of the text. This contradiction between the logic of image and text exists as long as one acknowledges that there is an opposition rather than a continuity between textual and visual languages. Baetens concludes by offering a solution in Anne-Marie Christin's concept of "screen thinking," in which distinctions between the visual and the textual are merely superficial. The fundamentally pictorial non-Western language systems and alphabets show how the Western schism of word and image is historically contingent, constructed and defined by Western dichotomies.

Wiley, looking at the reception of the image from the perspectives of cognitive science, presents a discussion of how to balance the benefits and costs of using image-rich presentations (including multimedia and virtual reality) for understanding, problem solving, and comprehension of specific content areas. Analyzing a broad range of applied research in cognitive science, she summarizes both the benefits and the dangers of images for perception and understanding, concluding that images and animations might in some cases actually decrease understanding of the texts they accompany, distracting the reader from understanding the central message of a particular text. The empirical research that she summarizes tests the use of "visually rich" presentations, visual adjuncts to texts, and multimedia and outlines a number of conditions that constrain what kinds of images are useful in facilitating comprehension and how such presentations should be structured so that they work with, rather than against, the text they are intended to support: for example, the differences between processing realistic images and processing symbolic or abstract illustrations, the importance of minimizing the competition between picture and text, and the usefulness of animations in helping learners visualize complex, interactive, multidimensional data. She concludes that students do not learn from "simple transmission of information" but must have contextually relevant visual information and be able to take an active and constructivist problem-solving approach to knowledge. By looking at how people construct knowledge, Wiley uses specific moments of interpretation to determine how effectively images convey meaning in those moments.

Each of the chapters in this section uses traditions of cognitive, developmental, and cultural theory to critique influential theories of communication and perception and offers a counterargument to the idea that visual knowledge invokes a communications revolution. All demonstrate that different ways of knowing, whether based on cognitive processes or cultural practices, help us understand how the use of images affects communication and often results in contradictory logics that, in turn, offer implications for new media.

Identities and Cultures in Digital Designs

Debates about the status of images in the history of communication systems and in the conventional interpretations of various texts lay important groundwork for the design of hybrid documents or presentations in new media environments. The volume's last section focuses on how new media artifacts construct hybrid experiences, identities, epistemologies, and virtual realities. Taken together, they provide an overview of some significant uses and implications of new media in a variety of interactive digital environments. These digital environments, ranging from the more familiar World Wide Web and interactive multimedia to immersive virtual reality environments, all offer opportunities to construct hybrid and playful identities and to compose visual and verbal representations of constructed texts, selves, and cultural spaces.

Gail Hawisher and Patricia Sullivan examine in chapter 10 the "cyberhybridity" of identities constructed by feminist sites on the Web. Using feminist theories of technology and geography, they point out how "'home' is culturally constructed as a web of place, community, gender, class, ethnic, institutional, disciplinary, and national affiliations" and how " home space" thus becomes "gendered, hierarchical, spatial, and contested" in its symbolic representation on the Web. Drawing on several previous studies of women online and then turning to examine international feminist Web sites, Hawisher and Sullivan argue that the women who develop and operate such Web sites construct hybrid cyborg cultural identities "to resist current cultural and social formations." These women use the Web to design gender and complex international identities, in one case, blurring their European and American cultures (one group are recent immigrants who live in the United States but construct a playful site about Russian female identity). All these women obviously participate in the dominant American economy and privilege, but, like the cyborg figure, their constructed hybridity allows them to create oppositional identities, to form alternative cultural narratives, and to bring women and capital together in their own interests, thus performing activism online. Hawisher and Sullivan conclude that the impact of new media on writing and learning reaches new heights that cannot be easily characterized by any traditional definitions of gender or national identity.

Alice Crawford takes communication and identity even farther away from our experiences of traditional texts or digital environments in chapter 11 when she suggests that immersive virtual reality (VR) technologies can be used to create multiple identities. Defining VR as an "immersive human-computer interface," she argues for the primary use of images in crafting self-identification. Taking issue both with traditional media critics who suggest that VR will "provoke forms of identification in which the 'glamour' (in the oldest sense of the word) of images on screen will work a bewitching spell on our psyches" and with progressive feminist critics like Donna Haraway whose cyborg identity asks to "break free from its moorings in human embodiment." Crawford instead advocates that we appropriate VR to enhance "our relations with our own embodied selves." Using insights from psychoanalytic theory, Crawford suggests that new visualization technologies like VR can offer both a complex and a holistic vision of the constructed self. VR can thus be used to create images that expand our notion of identity in socially positive ways and can have a positive affect on our understanding of differences in and through embodied identity. Crawford establishes the real potentials and visual possibilities of the immersive environments VR can offer by calling for an "ethic of spectatorship that values such experiences and incorporates entertainment, escapism, and play in a collective search for the good life," citing the general availability of authoring tools to create a wide range of avatars, or visual representations of the self. This fantasy of representation, she argues, offers a momentary escape that can enhance our embodied relations.

In an innovative approach to problem-based learning, Ellen Strain and Gregory VanHoosier-Carey describe in chapter 12 the theory and design of their interactive application, *Griffith in Context*, which looks at the cultural, formal, and historical features of the film *The Birth of a Nation* and engages the user in the specific construction of segments and interpretations of the film. Strain and VanHoosier-Carey first demonstrate how multimedia is particularly suited to a humanities scholar's ability to "link together a vast array of primary texts (written, visual, cinematic, material), secondary texts (historical studies, critical articles), and lived practices" and his or her recognition of patterns and associations among cultural artifacts into a dynamic form. Arguing that design practices are rooted in Aristotelian ideas about art and rhetoric, they assert that humanities computing applications will increasingly demonstrate the rhetorical power of interfaces. Those cultural values present in both the film and the interface are laid bare, showing their own construction, and are highlighted by an interactive project design that incorporates navigable filmstrips, the spectator's experience of narrative immersion, scholarly voice-over commentary, and color-coded "indices of analysis" (Historical Re-creation, Racial Representation, Filmic Technique, and Literary Origins). The interactive interface they designed for *Griffith in Context* thus provides valuable lessons about critique and composition from the constructed and hybrid nature—what many refer to as the "hypermediated structure"—of new media. By avoiding the temptation to adapt their project to traditional modes of scholarly argument, they highlight the value of multimedia design as enabling constructivist learning and dialogic approaches to knowledge.

In the volume's concluding chapter, Josephine Anstey takes this kind of experiential knowledge and moves to the far edges of "writing" with her fascinating account of her work programming, designing, and writing interactive fiction in immersive virtual reality, creating an experience that "can plunge the reader-user into a three-dimensional audiovisual world that responds in real time to her interventions." Anstey's project, *The Thing Growing*, originally built for a CAVE VR system in a room-sized, projection-based, virtual reality theater, features an interactive "creature" she calls the Thing that taunts and challenges the user. Anstey's account of creating the Thing blurs the relationship between fictive, real, and various embodied identities as she pushes us to consider the meshing of narrative and the visual. Her description of the design process highlights how graphics in the CAVE must be created from the user's point of view, changing as she moves, using cuts and timing in ways both similar to and different from film conventions. More importantly, the "realism" of the experience lies not in the literal re-creation of particular objects and spaces, but in the character's responses and in the psychological impact, both of which are constrained by the narrative script. Anstey demonstrates the hybrid experience of new media by showing how both visual perception and textual narrative conventions emerge as the basis for engendering emotions and our experience of "identities" and "reality."

In a sense, all the authors in this final section, whether looking at constructed online identities or the promises of digital media designs, suggest that the rich history of both theories of the visual and theories of the textual contribute to production but recognize that these theories cannot capture the intricate dialogic processes and hybrid identities of new media practice and experiences. Such a recognition reminds us of how Michel de Certeau (1984) defines methods of practice. Practices, he says, must occur in specific, culturally controlled contexts, but they also often exceed and complicate those contexts in surprising ways:

> [J]ust as in literature one differentiates "styles" or ways of writing, one can distinguish "ways of operating"—ways of walking, reading, producing, speaking, etc. These styles of action intervene in a field, which regulates them at a first level . . . , but they introduce into it a way of turning it to their advantage that obeys other rules and constitutes something like a second level interwoven into the first. . . . [T]hese ways of operating . . . create a certain play in the machine through a stratification of different and interfering kinds of functioning. (48)

The rules, dictates, and history of both visual literacy and textual literacy come into play as "styles of action" on what de Certeau would call the "first level"; however, our conscious practices, our "ways of operating," our ways of producing and understanding new media produce a second level "interwoven into the first." This metaphor resonates nicely with the discussion of new media design presented here because, as each chapter in this volume outlines specific forms of design and practice, it delineates part of that complex stratification of functioning that constitutes new media's "play" in our digital machines. By looking at new media theories and instances of practice within the stratified, conflicting networks of interpretation, the authors in this volume present important new ways to be nonmodern. Specific instances can move us beyond the merely theoretical to interpret and to create with a fully hybrid eloquence, and the examples offered in the chapters that follow become those everyday practices that enact the verbal and visual complexity of new media.

Notes

1. Stuart Moulthrop's 1991 essay, "You Say You Want a Revolution? Hypertext and the Laws of Media" remains a classic argument for defining hypertext as a "new" form of media by applying McLuhan's laws, and Jay David Bolter (1991) explored continuities and changes in the history of writing, identifying the late twentieth century as the "late age of print" in *Writing Space.* Scholarship in hypertext theory and in computers and composition has sustained an ongoing inquiry into how significantly digital on-line environments challenge or blend together multiple modes of literacy practice, often resulting in transformed pedagogy that requires what the New London group educational theorists call "multiliteracies": educational practices that emphasize design for social change (Kress 1998, 56–57). For other examples of these changes in literacy practices, see Heba 1997; Hocks 1999; Joyce 1995; Kaplan 1999; Kress 2000; Lanham 1993; Moulthrop 1994; Murray 1997; Selfe 1999; Snyder 1998; Taylor and Ward 1998; and Tuman 1992.
2. For other critiques of new media technology as revolutionary, see especially Aarseth 1997; Bolter and Grusin 1999; Hocks 1995; Kendrick 2001; and Terry and Calvert 1997.
3. See especially Coyne 1995; Kress 1998; Mitchell 1992; Schriver 1997; Sullivan 2001; Tufte 1983, 1997; and Wysocki 2001 as works by scholars who have stressed the fully visual nature of how information and communication technologies construct knowledge.

Works Cited

Aarseth, Espen J. 1997. *Cybertext: Perspectives on Ergodic Literature.* Baltimore: Johns Hopkins University Press.

Birkerts, Sven. 1995. *The Gutenberg Elegies: The Fate of Reading in the Electronic Age.* Winchester, MA: Faber and Faber.

Bolter, Jay David. 1991. *Writing Space: The Computer, Hypertext and the History of Writing.* Hillsdale, NJ: Lawrence Erlbaum.

Bolter, Jay David, and Richard Grusin. 1999. *Remediation: Understanding New Media.* Cambridge, MA: MIT Press.

Coyne, Richard. 1995. *Designing Information Technology in the Postmodern Age.* Cambridge: MIT Press.

de Certeau, Michel. 1984. *The Practice of Everyday Life,* trans. Steven Rendall. Berkeley and Los Angeles: University of California Press.

Dondis, Donis A. 1973. *A Primer of Visual Literacy.* Cambridge: MIT Press.

Heba, Gary. 1997. "HyperRhetoric: Multimedia, Literacy, and the Future of Composition." *Computers and Composition* 14, no. 1 (January):19–44.

Hocks, Mary E. 1995. "Technoptropes of Liberation." *Pre/Text* 16, nos. 1–2 (Spring/Summer):98–108.

Hocks, Mary E. 1999. "Toward a Visual Electronic Critical Literacy." *Works and Days* 17, nos. 1–2 (Spring/Fall):157–172.

Joyce, Michael. 1995. *Of Two Minds: Hypertext Pedagogy and Poetics.* Ann Arbor: University of Michigan Press.

Kaplan, Nancy. 1999. "E-Literacies: Politexts, Hypertexts, and Other Cultural Formations in the Late Age of Print." Online. Available: <http://raven.ubalt.edu/staff/Kaplan/lit/One_Beginning_417.html> (accessed July 2, 1999).

Kendrick, Michelle. 2001. "Interactive Technology and the Remediation of the Subject of Writing." *Configuations: A Journal of Literature, Science and Technology.* 9, no. 2 (Spring):231–251.

Kress, Gunther. 1998. "Visual and Verbal Modes of Representation in Electronically Mediated Communication: The Potentials of New Forms of Text." In *Page to Screen: Taking Literacy into the Electronic Era,* eds. Ilana Snyder and Michael Joyce, 53–79. London: Routledge.

Kress, Gunter. 2000. "Multimodality." In *Multiliteracies: Literacy Learning and the Design of Social Futures,* ed. Bill Cope and Mary Kalantzis, 182–202. New York: Routledge.

Landow, George P. 1992. *Hypertext: The Convergence of Contemporary Critical Theory and Technology.* Baltimore: Johns Hopkins University Press.

Lanham, Richard A. 1993. *The Electronic Word: Democracy, Technology and the Arts.* Chicago: University of Chicago Press.

Latour, Bruno. 1993. *We Have Never Been Modern.* Cambridge: Harvard University Press.

McCorduck, Pamela. 1995. "How We Knew, How We Know, How We Will Know." In *Literacy Online: The Promise (and Peril) of Reading and Writing with Computers,* ed. Myron C. Tuman, 245–259. Pittsburgh: University of Pittsburgh Press.

Mirzoeff, Nicholas, ed. 1998. *The Visual Culture Reader.* New York: Routledge.

Mitchell, William J. 1992. *The Reconfigured Eye: Visual Truth in the Post Photographic Era.* Cambridge: MIT Press.

Mitchell, W. J. T. 1995. *Picture Theory: Essays on Verbal and Visual Representation.* Chicago: University of Chicago Press.

Moulthrop, Stuart. 1991. "You Say You Want a Revolution? Hypertext and the Laws of Media." *Postmodern Culture* 1, no. 3 (May). Online. Available: <http://www.iath.virginia.edu/pmc/text-only/issue.591/moulthro.591>.

Moulthrop, Stuart. 1994. "Rhizome and Resistance: Hypertext and the Dreams of a New Culture." In *Hyper/Text/Theory*, ed. G. P. Landow, 299–322. Baltimore: Johns Hopkins University Press.

Murray, Janet H. 1997. *Hamlet on the Holodeck: The Future of Narrative in Cyberspace.* Cambridge: MIT Press.

Ong, Walter. 1982. *Orality and Literacy: The Technologizing of the Word.* New Accents Series, gen. ed. Terrence Hawkes. New York: Routledge.

Postman, Neil. 1992. *Technopoly: The Surrender of Culture to Technology.* New York: Knopf.

Schriver, Karen E. 1997. *Dynamics in Document Design.* New York: Wiley.

Selfe, Cynthia L. 1999. "Technology and Literacy: A Story about the Perils of Not Paying Attention." *College Composition and Communication* 50, no. 3 (February):411–436.

Snyder, Ilana, ed. 1998. *Page to Screen: Taking Literacy into the Electronic Era.* London: Routledge.

Sullivan, Patricia. 2001. "Practicing Safe Visual Rhetoric on the World Wide Web." *Computers and Composition* 18, no. 2:103–122.

Taylor, Todd, and Irene Ward, eds. 1998. *Literacy Theory in the Age of the Internet.* New York: Columbia University Press.

Terry, Jennifer, and Melodie Calvert. 1997. *Processed Lives: Gender and Technology in Everyday Life.* New York: Routledge.

Tufte, Edward. 1983. *The Visual Display of Quantitative Information.* Cheshire, CT: Graphics Press.

Tufte, Edward. 1997. *Visual Explanations: Images and Quantities, Evidence and Narrative.* Cheshire, CT: Graphics Press.

Tuman, Myron. 1992. *Wordperfect: Literacy in the Computer Age.* Pittsburgh: University of Pittsburgh Press.

Wysocki, Anne Frances. 2001. "Impossibly Distinct: On Form/Content and Word/Image in Two Pieces of Computer-Based Interactive Multimedia." *Computers and Composition* 18, no. 2:209–234.

I

Visual and Verbal Practices in New Media

| 1 |
CRITICAL THEORY AND THE CHALLENGE OF NEW MEDIA

Jay David Bolter

The Verbal and the Visual

It is a commonplace to observe that we are living in an age dominated by visual representation, a commonplace shared by such diverse critics as E. H. Gombrich (1982, 137), W. J. T. Mitchell (1994, 2–3), and Frederic Jameson (1991, 299). In fact, this belief seems to have been shared by both popular culture and the art community for much of the twentieth century. During earlier centuries, in which print was our most prestigious medium, the balance between verbal and visual representation strongly favored the verbal. European and later North American culture chose to exploit the printing press to establish this unequal relationship. In most printed books, especially prior to the development of photolithography, words contained and constrained images. The images served as illustrations of the text, and the real work of communication was thought to be done by the words (Bolter 2001). With the development of a series of audiovisual technologies, however, beginning with photography and photolithographic printing and including film and television, the balance between word and image shifted.

Influenced by modernist art, graphic artists associated with the Bauhaus, de Stijl, and the Swiss or International Style reversed the accepted relationship by subsuming words into images and so teaching us to regard words themselves as images. These designers brought the modernist view into popular culture, by surrounding us in posters and magazine ads with examples of the word-as-image (Meggs 1998; Drucker 1994). Their work made the word immediate and sensually apprehensible by insisting on its visual form rather than its symbolic significance. Meanwhile, film, which was often said to be the preeminent popular art form of twentieth century, refashioned narrative forms and repurposed individual stories that had belonged to the novel and the stage play. Because they were such vivid audiovisual experiences, films seemed to offer greater immediacy and authenticity than novels or plays. Later television added its own definition of immediacy through its claim of "liveness." A television broadcast seemed to deliver images instantaneously to viewers throughout the country and eventually, with the aid of communication satellites, throughout the world. Unlike a newspaper or news magazine, a televised newscast could report events "as they happened." Television seems live, even when prerecorded material is being broadcast (Bolter and Grusin 1999).

Digital media continue in this line of challenges to the dominance of the printed word, by claiming to provide a new kind of interaction between the user/viewer and the digital application. Enthusiasts for digital media—from hypertexts and Web sites to stand-alone multimedia applications and simulations—insist that these new media forms can interact with the user in novel ways. The reader of a printed novel is expected to move in linear order from first page to last. When she visits a Web site, however, she points and clicks to determine which pages she will read and in which order. Thus, we think of the World Wide Web as a multilinear writing space. In fact, many genres of print are being refashioned for presentation on the Web: There are newspaper Web sites, textual repositories, political Web sites that fulfill the functions of campaign brochures and position papers, textual repositories of "classic literature," online encyclopedias, and so on. All of these sites both borrow from and seek to improve on their printed counterparts by promoting ease of access, interactivity, or flexibility as the advantages that the Web offers over print.

Designers for the Web create fluid combinations of words and images, which are both after all only raster-scanned pixels. Perhaps because words are not as easy to read on a conventional video screen as they are in print, Web designers often give more space and visual weight to images. In addition, the Web supports media forms not available to print. Streaming audio and video are already important on the Web, and their influence will increase as increasing bandwidth to the home improves the quality of such media streams. The Web can in fact absorb many of the media forms of the twentieth century. The Web's eclectic character also means that it can borrow or parody many of the relationships between word and image that have characterized earlier forms.

The techniques developed by graphic designers for printed magazines and billboards are now featured extensively on the Web: The treatment of the word as a static image is therefore common. Through streaming media the Web is now refashioning film and television as well. Film and television usually replace the word rather than subsuming it into the image. But Web designers still seem to prefer a different strategy, the one that they have borrowed from graphic design. Rather than replace words altogether, they refigure them as images or displace them by moving them to the margins in the act of communication. Web sites still include words used in their traditional symbolic role; there are after all millions of words on the Web. But the Web places static and moving images alongside words and asserts their equal and ultimately perhaps superior status. There is a danger that the words will not be able to compete with images, with their promise of immediate communication.

In short, the World Wide Web and other new media challenge not only the form of the book, but also the representational power of the printed word. Web users and traditional readers both understand this challenge at some intuitive level, and they may react with enthusiasm or despair, just as audiences did to the earlier challenges that occurred with film and television. For defenders of the traditional book, the Web may pose an even greater threat than television and film, for it was hard to imagine that our culture would ever seek to replace all written communication with those audiovisual media. The Web and other new media appear to accommodate both the verbal and the audiovisual (although they favor the audiovisual), so that enthusiasts do dare to suggest that the Web and other new media forms could replace books or libraries altogether (Kurzweil 1999).

Theory and Print

The challenge of new media has ramifications in many arenas in which print remains important to our culture. The academy is clearly one such arena, and in particular, both education and research in the humanities. This challenge is probably felt more keenly in the humanities than in the sciences. Although, as Elizabeth Eisenstein (1983) showed, modern science grew up in the media environment offered by the new technology of print, scientists now seem to regard print as simply a vessel in which their ideas can be transmitted. Other media forms might do as well. Scientists have certainly been enthusiastic in taking up electronic forms of transmission. It is worth remembering that Tim Berners-Lee created the World Wide Web to serve as a forum in which physicists and other scientists could share experimental results and drafts of papers. Those in the humanities are more committed to the printed book as their media form. Despite the efforts of cultural studies (and of course film studies and art history) to move beyond print, books and other printed materials still constitute the principal objects of study for most humanists. Almost all literature comes to humanists in print, even if the printed version may sometimes be a transcription of an originally oral work or a work from an earlier technology. Humanists prefer to continue to publish their critiques in the same form. This preference remains the same for radical theorists and for literary conservatives.

In fact, in challenging the status of print, visual digital media also call into question the status of critical theory in the academy. For decades, theory has been regarded as the most prestigious and intellectually rewarding work in the humanities. In English departments, as we know, those who teach theory enjoy a favored status in comparison with teachers in writing programs, who are concerned with writing practice. Yet critical theory remains grounded in the forms and practices of print technology. In the popular media, literary theorists are held up as the most radical members of our profession, either because they are aiming at a radical revision of our understanding of literature or simply because as neoMarxists they insist on grounding social change in cultural theory.

On the other hand, the attitude of cultural and literary theorists toward new media is itself often conservative and predictable. When they notice new media at all, the theorists' first impulse is to critique these new forms as implicated in the political economy of late global capitalism. For purposes of critique, they often regard new media forms as mass media, which they can analyze in terms provided by television, radio, magazines, and film. Like their predecessors, these new digital forms are the products of the political economy of late global capitalism—hence the focus on e-commerce on and off the World Wide Web. Examples include Andrew Ross's 1997 Internet article "The Great Wire Way" on alienation in Silicon Alley and historian David Noble's (1998a, 1998b) "Digital Diploma Mills," a critique of the rush by universities to make money on Web-based education. When cultural critics focus on the efforts of media companies to control the flow of information and entertainment to the developing world, they include digital media, such as the Web, applications software, and computer games, along with film and television in this analysis.

The cultural studies of new media theory must be critical, so long as cultural theorists regard new media as the latest extension of the "culture industry" identified decades ago by Theodor Adorno and Max Horkheimer (1993). Furthermore, if new media is the latest manifestation of late global capitalism, then the cultural critic will obviously want to maintain a critical distance from new media forms. If these forms are indelibly marked by capitalist ideology, then to become an active participant in these new media forms—to design Web sites or create multimedia applications—is to become complicit in that ideology. On the other hand, at least some cultural critics evidence an interesting ambivalence: They seem to want to critique and to participate at the same time. In their theoretical writings this ambivalence can be represented as the dissolving of false dichotomies. Donna Haraway's (1991) notion of the cyborg is influential in new media studies precisely because Haraway offers the cyborg as an opportunity to have it both ways: to provide a neoMarxist critique of the ideologies embodied in new media forms and yet to suggest at the same time that technoculture may be redeemed. To some extent, this ambivalence is a feature of the cultural studies of other popular media forms on television, in films, in magazines, and elsewhere. Although cultural critics recognize the sexism, racism, or classism in these forms, they continue to study them in detail. They study them precisely in order to explore the sexism, racism, or classicism, but they must nevertheless find something compelling about these forms, to which they devote months or years of scholarly effort. Often, they identify modes of resistance practiced by the audiences for these forms; this "resistant reading" can in a sense validate the media form.

New media differs, however, from traditional forms of mass media. Unlike television, radio, film, and popular magazines, new media are not always broadcast or presented to an audience of passive consumers. It is much easier to become a producer of a Web site or even a multimedia application than it is to publish a book or magazine or to produce a film or a television show. Although there are indeed mass Web sites such as amazon.com or Yahoo!, there are millions of small or individual sites as well, each of which is in theory equally accessible to Web users. Millions of people can and do participate in the framing or designing of their own commercial or private sites. These new media forms are available to us as producers as well as consumers, and they are available as forms of production to cultural critics and academics in general. Critics can put their essays up on the Web, and they sometimes do. *Postmodern Culture* <jefferson.village.virginia.edu/pmc> is a journal that publishes exclusively on the Web, and Andrew Ross's (1997) essay critiquing Silicon Alley is available on a Web site. But these are exceptions. Although individual essays and drafts circulate on the Web, in accord with Berners-Lee's original vision, what cultural critic would publish a book-length work exclusively in this form? Furthermore, the Web publications are usually simply electronic copies of linear essays intended for print; this is true even for the contributions to *Postmodern Culture*. The radical critique of new media is carried out in the traditional forms, essays and monographs for print. There may be changes in the style of presentation. Cultural studies and feminist studies have explored, for example, first-person accounts, reflecting the notion that the author should acknowledge the ways in which she enters into the cultural matrix. In other respects—linear presentation, discursive argumentation, the use of footnotes and references as scholarly documentation—the work of cultural studies and other theorists is traditional.

Although it remains grounded in the forms and practices of print technology, critical cultural theory has also enjoyed an ambivalent relationship to print. With the work of the poststructuralists from the late 1960s to the 1980s, theorists forged a critique of language and writing that was also implicitly a critique of print. As hypertext theorists have pointed out, the work of the poststructuralists undermined assumptions about the transparency and fixity of writing that were supported by the technology of printing (Landow 1997). Yet even as they deconstructed the book, the poststructuralists continued to produce printed books and essays, most of them conventional in form, if unorthodox in style. Critical theory needed the technology of print in order to deconstruct print. The same ironic relationship is true for the discourses that replaced poststructuralism as dominant at the end of the twentieth century. Cultural critiques of such popular media as magazines, television, and radio were aimed at increasing the legitimacy of the study of such media. As Grossberg, Nelson, and Treichler (1992) explain, "unlike traditional humanism [cultural studies] rejects the exclusive equation of culture with high culture and argues that all forms of cultural production need to be studied in relation to other cultural practices and to social and historical structures" (4). Although theorists accepted the cultural significance of popular media, they nevertheless continued to publish in print. Their traditional essays were really ekphrastic responses to popular forms, ekphrastic in the sense that they sought to describe and critique in words the visual images (or sometimes the music) that constituted the object of their research.

There is a tradition in humanistic studies of translating other media forms back into the medium of print, and this tradition continues with new media. Multimedia performances informed by theory are becoming more common, but their status within the academy is still problematic, as is shown by the debate over whether staging such performances should be given as much weight as papers and books in tenure decisions. Such performances are more likely to be the work of digital artists with a tangential relationship to the academy. The way to make a critical theoretical statement is still through an essay whose jargon may be new, but whose form has hardly changed in decades.

It is not that there is some inadequacy in printed media forms that digital forms can remedy: New digital media obviously have no claim to inherent superiority. Every media form, including print, is defined and at the same time constrained by the cultural practices that have accrued around that form. The printed essay and monograph are contextualized in this way. And because the contexts are shifting with the advent of visual (and perceptual) digital media, these printed forms can no longer guarantee to cultural theorists a privileged place from which to evaluate other media. When they use the standard scholarly forms of print, media critics are committing themselves to a particular perspective, in which the word is the privileged mode of representation and images are secondary and subsidiary. Yet our culture seems to be choosing a different valuation of word and image in its new media forms.

The gap between media theory and the cultural practices that surround new media forms has grown wide, perhaps so wide that theory can no longer effectively critique these new practices. To be effective—that is, to affect the choices that our culture is making—media theory needs to engage with the practice of digital media. Among humanists, it is teachers of writing who are actively seeking to close this gap between theory and practice.

Theory and Practice

Whereas literary and cultural theory continues to be written in a scholarly form that dates from the early twentieth century, teachers and scholars of writing and rhetoric have adopted the new electronic technology much more readily than their literary critical or cultural critical colleagues. Their readiness can be interpreted as a practical response to the realities of the business and bureaucratic worlds in which most writing is now done. If business and technical communication is increasingly electronic, then one must teach students to use the new digital tools available for writing. The fields of technical communication and computers and writing have an interest in electronic writing, however, that is something more than pragmatism. These scholars are neither conservative nor ignorant of cultural theory. Many of these scholars read and contribute to the same theoretical debates as cultural theorists (see Wysocki [ch. 2], Hawisher and Sullivan [ch. 10], and Strain and VanHoosier-Carey [ch. 12] in this volume). They hold many of the same radical positions with regard to the economic and ideological aspects of contemporary culture. They may combine these views, however, with a practical interest in electronic technologies of communication, which they believe they can use to radicalize their students.

So, for example, three such educators (Myers, Hammett, and McKillop 1998) can write: "We define *critical literacy* as the intentional subversion of meanings in order to critique the underlying ideologies and relations of power that support particular interpretations of a text. . . . Hypermedia is a particularly powerful environment for this critical literacy practice. . . . [H]ypermedia authoring can support the emancipation of one's self and others through the authoring and publication of critical texts that by questioning representations of the self, expand the possibilities for the self in future actions as a member of a community" (64–65). For such writing educators, a radical position at the theoretical level goes along with a radical practice. They are willing to engage with the technology, and this willingness perhaps renders them suspect to the theory community at large. Although publishing a linear essay on the Web is not suspect, creating a hypermedia artifact may be, precisely because it involves media forms that cultural theorists have come to associate with corporate software and entertainment giants.

This suspicion is just the most recent version of the tension between theory and practice that has been present in a variety of disciplines in the humanities over many decades. We can find this tension in music departments (between performers and musicologists), language departments (between theorists and teachers of the foreign language), and radio, television, and film departments (between communications or film theorists and practitioners), as well as English departments. We might ask whether there has ever been a successful synthesis of theory and practice in any of these disciplines. The solution has always been to keep the divisions as separate as possible (both within the structure of the university and as disciplines with their own journals and conferences) or simply to create a hierarchy within a single department. One irony is that the popular perception of the relative value of theory and practice often differs from the attitude within academic departments themselves. Outside of the university, for example, no one reads or cares much about film theory, whereas the practical film schools have produced some of the powerful and well-known people working in the entertainment industry today. Likewise, most legislators, who control the budgets of state universities, no doubt believe that teaching English composition is vastly more appropriate and important than the critique of cultural forms.

The attitude of American society toward education has historically been quite pragmatic: Education is supposed to lead to economic prosperity, as politicians still argue without any hesitation. So it is not surprising that outside the academy, practice should be valued above theory. This is true of new media studies as well as older disciplines. At the secondary and primary school levels, there has been a widespread, even enthusiastic acceptance of the educational possibilities of new media. Politicians and parents seem to agree that computers belong in schools and that schools should be connected to the Internet, which is all the more remarkable because there is otherwise so little social consensus about the future of American education. The computer and the Internet are coming to be part of the general educational experience. Such a privilege was not accorded to any of the earlier audiovisual media: Photography, radio, film, and television were all relegated a limited and specialized place in the curriculum. Yet our culture seems to regard the Internet and the Web as appropriate companions and in some cases replacements for printed texts and other educational resources. Although some educational theorists express concern about the public's unquestioning belief in the usefulness of the new technology, the wiring of American education continues.

In short, communication through new media is becoming a widespread practice in schools, within the university community, and in the worlds of business and bureaucracy after the university. Surrounded as they are by new media practice, humanist theorists might feel the need to engage themselves with new media as well. What would be the cost of such an engagement? How might theory make itself useful to practice? It seems clear that to do so, theory would have to change its demand for critical distance. Theorists would have to follow the lead of the writing community and engage in new media production themselves in order to provide useful insights. This does not mean that theorists would have to become unthinking enthusiasts for technology. Teachers in the graphic and fine arts move easily back and forth between practice and critique. They produce their own artifacts, but they also criticize the artifacts and "design problems" of their students. Theorists might learn to imitate this model and find themselves oscillating between the two modes of making and critiquing. To do so, however, theorists would have to appreciate the shifting balance of the verbal and the audiovisual modes of representation, to which new media are contributing. To approach new media as a practice is to appreciate the cultural significance of images and sounds as well as written words.

Although such an approach to practice would require significant changes for critical theory, it would not require the abandoning of the notion of critique itself. The writing community has shown that critical theory can inform the educational use of Multi-user Object-Oriented environment (MOOs), chatrooms, and similar environments. It has shown, for example, how these environments can become sites for exploring the meaning of gender and race in postmodern identity. Although MOOs and chatrooms have formal technological properties that students must master, these properties in no way free these environments from the cultural contexts in which they are embedded. Instead, the properties of MOOs and chatrooms give them a special cultural valence. Cyberspace remains part of our physical and social world, and at least for wealthy industrial societies, it is becoming a key place for the definition of postmodern identity.

An understanding of how technologies such as MOOs and chatrooms grow simultaneously out of and into their culture is precisely what critical theory can contribute to the evolving practice of new media. Critical theory can contextualize practice. To do so, however, theory must be framed in such a way as to inform or reform practice. This means that theory may need to reform itself in order to speak to a practice that is visual and aural (perceptual in general) as well as textual. As we have seen, theory has been print-based and therefore embodied in a particular set of media forms (e.g., the critical journal article, the anthology essay, and the monograph). The primacy of these forms is called into question by the ways in which our culture is using new media, so that the printed article and monograph now have to confront competing forms on the World Wide Web and in DVD. In this way the practice of new media challenges theory, and the challenge comes in large part from the change in the ratio between the verbal and the visual that characterizes new media as compared to traditional forms. With its work in MOOs and discussion groups, the field of computers and writing has shown the way toward a theory that can be expressed in new media practice (see Haynes and Holmevik

1998). But precisely because of their commitment to such text-based applications, these scholars have not yet fully addressed all aspects of the new visual and perceptual media.

Design in Context

At Georgia Tech, both in our Wesley Center for New Media Education and Research and in our graduate program in information design and technology, we are exploring the interplay of media theory and practice, while at the same time seeking to understand the changed relationship between the verbal and visual that is emerging in new media forms. We believe that an engaged media theory can make an important contribution to practice by elucidating the contexts for design. In this spirit we apply cultural and historical theory to inform and reform the making of Web sites, stand-alone multimedia, and mixed-media performances and installations. Our faculty draw on their training in narratology, film theory, performance theory, and cultural and media studies to provide contexts for their designs. These designs embody critique, but the critique speaks through the artifacts themselves, rather than over against the artifacts.

The history of media provides one important critical context for design. My colleague Richard Grusin and I have argued that new media forms define themselves by borrowing from and refashioning earlier forms in print, photography, radio, film, and television. We have called this process of borrowing and refashioning "remediation" to emphasize that new media forms always claim to be improving or reforming earlier forms, even when they are paying homage to those forms by borrowing from them. What they borrow is a sense of the real or the immediate, and at the same they try to provide the user with a greater sense of immediacy (Bolter and Grusin 1999). We have argued that remediation has been going on for centuries as media forms have been introduced into our culture and entered into competition with existing media. Other colleagues in our department are pursuing similar historical investigations into the contested relationships between new media and earlier forms. Ellen Strain (1999) has explored the representation of virtual reality in contemporary film. Paul Young (1999) has explored the "anxiety" that film is experiencing as it is confronted with computer graphic technologies.

How can an awareness of such historical contexts inform new media design? In teaching Web design, I have used the notion of remediation to provide structure for students in the process of design. I offer the notion of remediation as a counterbalance to the rhetoric of revolution that is so common among new media enthusiasts, who insist that new media artifacts are, or should be, utterly new and original (e.g., Holtzman 1997, 15). Their rhetoric derives ultimately from high modernism, which assumed that to be creative one must break completely with tradition and formulate new first principles (Bolter and Grusin 1999, 49–50). Students often succumb to this rhetoric and use it as an excuse for ignoring traditions and techniques from other media forms. When we examine the Web, we do not find this supposed utter originality, but rather numerous remediations of such earlier media forms as newspapers and television news broadcasts, corporate brochures, mail-in forms, soap operas, encyclopedias, top-40 radio stations, and printed and video pornography. Thus, remediation is fundamental to Web design. To paraphrase the modernist T. S. Eliot, all Web designers repurpose; good Web designers remediate. To repurpose is simply to pour content from one media form into another, while attempting to replicate the earlier medium's definition of the authentic. To remediate is to borrow the sense of the authentic from one media form and to refashion it for another. Whereas accomplished Web designers do this intuitively, I have asked my students to make the process explicit, by choosing a printed form and consciously refashioning it for the Web. One group of students refashioned some of the formal characteristics of the comic book; their goal was to use a subtle degree of interactivity and animation without abandoning the underlying static graphic form of the printed version. Another group designed a Web site as a critical refashioning of *YM*, a fashion magazine for teenage girls. Their project showed that remediation is not limited to formal qualities. New media applications can borrow the social meanings of earlier media and may choose simply to reproduce those meanings ("respectful remediation") or to reform them ("critical remediation"). The *YM* Web site was an explicit critique of the commodification of beauty that characterized the printed magazine.

Other faculty members in our program have adopted different historical approaches. In *Hamlet on the Holodeck* (1997), Janet Murray explores the significance of computer technology in the history of narrative media form and argues for both historical continuity and innovation. On the one hand, she shows how the new medium extends a tradition that includes such earlier forms as the Victorian novel, Elizabethan tragedy, and even Homeric storytelling. At the same time, she proposes a series of qualities that make this new narrative medium unique: new forms of immersion, agency, and transformation that grow out of the procedural character of computer-controlled narrative. Murray has examined these qualities in her work on interactive narrative with students at MIT and now at Georgia Tech.

Historical media studies and cultural studies come together in the design work of Gregory vanHoosier-Carey and Ellen Strain, especially in their pedagogical project *Griffith in Context.* (2000; see also this volume). This stand-alone multimedia application introduces students to the complex of formal and ideological issues that cluster around the film *The Birth of a Nation*. The application offers the students segments of the film together with excerpts from the novels *The Clansman* and *The Leopard's Spots* as well as other historical sources and views of contemporary film scholars. Through hands-on exercises, students come to appreciate the formal innovations in Griffith's film; at the same time they are encouraged to examine the racist ideology that infuses the film. *Griffith in Context* confronts the students with the ironic juxtaposition of Griffith as formal innovator and as revisionist social commentator. Film theory and cultural studies are the perspectives from which this irony is explored; these perspectives not only constitute the theoretical background but are also engrained in the interface of the application itself.

Still other disciplines have provided the critical contexts for the work on distance learning and electronic communication of Tyanna Herrington and Peter McGuire. Herrington applies the perspectives of rhetoric and communications theory to her Global Classroom Project, in which a set of asynchronous electronic tools links students in classes taught simultaneously in Atlanta and in St. Petersburg, Russia. The students consider the problems of international communication while at the same time experiencing those problems as they use electronic environments to bridge the spatial, temporal, and cultural gaps that separate them. McGuire has confronted similar issues in his remediation of a Georgia Tech course on the history of science fiction. First taught in the conventional classroom format, the course was converted to the now traditional version of distance learning: a series of videotaped lectures. McGuire has created a hybrid course consisting of both face-to-face meetings and Web materials, including the lectures, delivered as streamed video. Three different levels of mediation are contained in the final product, allowing McGuire to test what communications and educational theory tell us about authenticity in the processing of learning.

All of the above examples of contextualized design are pedagogical. Meanwhile, Diane Gromala and Sha Xin Wei are showing how theory can inform digital art and performance. Working in the long tradition of art as cultural critique, they explore the critical potential and cultural valence of particular new media forms, such as live digital performance and virtual reality.

Sha (2002) is examining what he calls "the architecture of responsive media spaces." His commitment to the marriage of theory and practice is clear in his description of the work of his Topological Media Lab (TML), which "provides a locus for studying gesture and materials from phenomenological, social and computational perspectives. TML research invents and evaluates dynamical structures that can support novel technologies of writing and the architecture of hybrid interactive spaces. The products of the laboratory are (1) scholarly presentations, (2) media artifacts and performances as pieces of cultural experiment, (3) opportunities for students of design to sharpen criti-

| **Figure 1.1** |

A dancer in the installation space T-Garden, created by Sponge <www.sponge.org>.

cal faculties in project-based work." Sha is also a member of Sponge, an international group of designers, artists, programmers and theorists who are building a digital architectural space called "T-Garden" (figure 1.1): Participants put on special clothes embedded with sensors and then walk, dance, and play in a space in which the visual forms and sounds are modified by their gestures. In an installation like T-Garden, the boundary between theory and practice dissolves: The practice is the theory, in this case a theory about the relationship of embodiment to digital technology.

Diane Gromala's [1995] work with virtual reality also dissolves boundaries, and she too is concerned with issues of embodiment. She began her Virtual Bodies project in the early 1990s but continues to work with students at Georgia Tech on the ideas that grew out of the project. Virtual Bodies takes as its task to dissolve or at least to explore the boundary between mind and body as it is represented in virtual reality (VR). Many cultural critics, including Allucquère Rosanne Stone (1991) and Donna Haraway (1991), have argued that VR represents yet another attempt by high technology to deny the significance of embodied ways of knowing, that VR belongs to the long Cartesian

| **Figure 1.2** |

Dancing with the Virtual Dervish, created by Diane Gromala and Yacov Sharir. Image courtesy of Diane Gromala.

tradition of privileging abstract thought over lived and felt experience. Gromala offers a nuanced critique of this argument. In her project she shows how VR can provide a unique, liminal, sensory experience, precisely because it can give us the sensation of disembodiment. Gromala's current project (figure 1.2) is a virtual tour of her body, involving 3-D graphics as well as various visual and verbal texts that the doctors produced in studying her condition, including X-rays, magnetic resonance imagings (MRIs), and technical reports. Wearing a headset the user is invited the read the "book" that contemporary medical practice has made of Gromala's body.

Gromala's Virtual Bodies project thus explores and partly dissolves not only the boundary between mind and body, but at the same time the boundary between text and image, the very boundary that the new media in general are calling into question. The Virtual Bodies project brings about these dissolutions not by discussing them in a discursive essay, but by performing them in a VR installation. The project is itself a practical critique of the view put forth by cultural studies theorists that VR is disembodying. The contexts for Virtual Bodies thus include both the popular reception of VR and the theoretical work throughout the 1980s and 1990s on identity, embodiment, and technology.

Why have we at Georgia Tech been able to take these first steps toward bridging the gap between theory and practice in new media? The answer lies in the creativity of faculty members who were committed to combining theory and practice even before they came to Georgia Tech. Ten years ago, our department changed its name from the English Department to the School of Literature, Communication, and Culture and instituted an undergraduate degree in the cultural studies of science and technology. New media studies grew out of that matrix. The school also been able to hire from a range of disciplines. In addition to literary theorists and film theorists we now have faculty with considerable experience in academic computing, the computer media industry, and digital art. Our new media faculty work closely with the Graphics, Visualization, and Usability Center in the College of Computing. The fact that we are located within a technical university has made it much easier to redefine our role to include practice as well as theory.

Because it depends on the specific institutional contexts at Georgia Tech, our department's shift may not serve as a model for the reconfiguration of humanities departments in the age of new media. In fact, there is not likely to be a single model for such reconfigurations. Many departments in universities and schools of art and design are now seeking to redefine themselves, and the process of redefinition depends on local politics as well as the vision of a few leaders. For example, the programs in Design | Media Arts at the UCLA School of the Visual Arts <www.design.ucla.edu/>, in Comparative Media Studies at MIT <web.mit.edu/21fms/www/>, in Media Studies at the University of Virginia <www.virginia.edu/topnews/releases2000/mediastudies-sept-19-2000.html>, in Interactive Telecommunications at the Tisch School of the Arts at NYU <www.nyu.edu/tisch/itp.html>—all are the result of interdisciplinary combinations that were appropriate for the particular institution in question. New media are eclectic, and we may expect that new media programs will emerge in an eclectic and even opportunistic fashion.

The Formal and the Ideological

As I have indicated, the work in electronic pedagogy and digital art at Georgia Tech depends on the combination of disciplines and perspectives of our particular faculty. What all our work has in common is an insistence that cultural and historical issues should not be separated from formal issues of design. For the tension between practical and critical theory for new media can also be understood as a tension between two modes of analysis, which we could call formal and ideological. What graphic designers and human-computer interaction (HCI) specialists appreciate in new media artifacts are the formal qualities. In the case of graphic design, these formal qualities are expressed in such terms as clarity, harmony, cohesiveness, and restraint. HCI makes formal measures of an artifact's "usability": how easily the user can perform tasks presented through the interface the artifact represents. On the other hand, contemporary critical theory is by nature suspicious of such analysis, which it associates with the modernist's emphasis on the timelessness of pure form. Explaining the formal characteristics of media in terms of their underlying ideologies, the critical theorist rejects the notion that these formal characteristics could be disinterested or universal in their application. So, for example, psychoanalytic film theory explained the apparatus of the cinema (the camera work, editing, the position of the spectator in the darkened theater) as expressions of masculinist or capitalist ideologies. Likewise, the critique of new media regards sophisticated, asymmetric Web design as yet another version of the fetishized image in American advertising and regards interactive computer games as a new genre of commodified violence.

These ideological readings, however, are not sufficient in themselves to explain new media. A new critical theory is needed that can make us aware of the cultural and historical contexts (and ideologies) without dismissing or downplaying the formal characteristics of new media. This theory needs to explain these formal characteristics without explaining them away, because practitioners have no choice: If they wish to create successful product, they must attend to these formal values (which used to be regarded as aesthetic values in art or utilitarian considerations in software engineering and computer programming). Any theory that is going to be useful for actual practice must offer the practitioner guidance in conceiving and executing the form of her work. A new critical theory should offer in addition an understanding of the cultural contexts in which the form is embedded. Such a theory should analyze and even criticize current cultural practices through new media forms. Instead of holding up new media forms such as the World Wide Web as examples of the excesses of late-capitalist culture, however, a new theory should turn new media forms themselves into vehicles of critique. Design in context must be critical and productive at the same time.

Throughout this chapter I have been exploring dichotomies: practice and theory, critique and production, the formal and the ideological, the visual and the verbal. These dichotomies can be thought of as aligned with one another:

theory	practice
critique	production
the ideological	the formal
the verbal	the visual

I am aware that these are all provisional dichotomies that must ultimately collapse. None of them can be or should be sustained in the university community, especially if universities are going to participate meaningfully in the cultural work of exploring the possibilities of new media. For the present, however, the dichotomies reflect real and important tensions between media theorists on one hand and designers and producers on the other. Theorists in media studies see it as their task to critique new media, as they have done with traditional broadcast media. They frame their critique in the language of ideological analysis, and they express the critique in linear essays intended for print. New media practitioners see it as their task to produce artifacts. They analyze those artifacts in formal terms, and in the current world of multimedia their artifacts emphasize visual or perceptual elements, at the expense of words. These dichotomies or tensions are keenly felt by both groups. How can we collapse them? The answer that we are pursuing at Georgia Tech is to explore the means by which the critical theory of new media can be expressed in and through new media artifacts themselves.

Acknowledgments

My thanks for the creative efforts of my students in two years of my Web design course, LCC 6111: especially, to Noel Moreno and Jennifor Gordon for the remediated comic book and to Chrissy Hess and Keith Freck for *YM*.

Works Cited

Adorno, Theodor, and Max Horkheimer. 1993. "The Culture Industry." In *The Cultural Studies Rreader.* ed. Simon During, 30–43. London: Routledge.

Bolter, Jay David. 2001. *Writing Space: The Computer, Hypertext, and the Remediation of Print.* Mahwah, NJ: Lawrence Erlbaum.

Bolter, Jay David, and Richard Grusin. 1999. *Remediation: Understanding New Media.* Cambridge: MIT Press.

Drucker, Johanna. 1994. *The Visible Word: Experimental Typography and Modern Art, 1909–1923.* Chicago: University of Chicago Press.

Eisenstein, Elizabeth. 1983. *The Printing Revolution in Early Modern Europe.* Cambridge: Cambridge University Press.

Gombrich, E. H. 1982. *The Image and the Eye: Further Studies in the Psychology of Pictorial Representation.* Ithaca, N.Y.: Cornell University Press.

Gromala, Diane. 1995. "Dancing with the Virtual Dervish: Virtual Bodies." In *Immersed in Technology: Art and Virtual Environments,* eds. Mary Anne Moser and Douglas MacLeod, 281–285. Cambridge: MIT Press.

Grossberg, Lawrence, Cary Nelson, and Paula Treichler, eds. 1992. *Cultural Studies.* New York: Routledge.

Haraway, Donna J. 1991. *Simians, Cyborgs and Women: The Reinvention of Nature.* New York: Routledge.

Haynes, Cynthia, and Jan Rune Holmevik, eds. 1998. *High Wired: On the Design, Use, and Theory of Educational MOOs.* Ann Arbor: University of Michigan Press.

Holtzman, Steven. 1997. *Digital Mosaics: The Aesthetics of Cyberspace.* New York: Simon & Schuster.

Jameson, Fredric. 1991. *Postmodernism, or, the Cultural Logic of Late Capitalism.* Durham: Duke University Press.

Kurzweil, Raymond. 1999. "The Future of Libraries." In *Cyberreader,* 2nd ed., ed. Victor Vitanza, 291–303. Needham Heights, MA: Allyn and Bacon.

Landow, George P. 1997. *Hypertext 2.0: The Convergence of Contemporary Critical Theory and Technology.* Baltimore: Johns Hopkins University Press.

Meggs, Philip. 1998. *A History of Graphic Design.* New York: Wiley.

Mitchell, W. J. T. 1994. *Picture Theory.* Chicago: University of Chicago Press.

Murray, Janet. 1997. *Hamlet on the Holodeck.* Cambridge, MA: MIT Press.

Myers, Jamie, Roberta Hammett, and Ann Margaret McKillop, 1998. "Opportunities for Critical Literacy and Pedagogy in Student-authored Hypermedia." In *Handbook of Literacy and Technology: Transformations in a Post-typographic World,* eds. David Reinking, Michael C. McKenna, Linda D. Labbo, and Ronald D. Kieffer, 63–78. Mahwah, NJ: Erlbaum.

Noble, David. 1998a. "Digital Diploma Mills, Part 1: The Automation of Higher Education." *October* (Fall): 107–117.

Noble, David. 1998b. "Digital Diploma Mills, Part 2: The Coming Battle over Online Instruction. *October* (Fall):118–129.

Ross, Andrew. 1997. The Great Wire Way. Available at <www.ljudmila.org/nettime/zkp4/28.htm>. Accessed on January 29, 2002.

Sha, Xin Wei. 2002. Available at <www.occ.gatech.edu/~xinwei/>. Accessed on January 29, 2002.

Stone, Allucquère Rosanne. 1991. "Will the Real Body Please Stand Up?" In *Cyberspace: First Steps,* ed. Michael Benedikt, 81–118. Cambridge: MIT Press.

Strain, Ellen. 1999. "Virtual VR." *Convergence* (Summer):10–15.

vanHoosier-Carey, Gregory, and Ellen Strain. 2000. *Griffith in Context.* CD-ROM. Unpublished.

Young, Paul. 1999. "The Negative Reinvention of Cinema: Late Hollywood in the Early Digital Age." *Convergence* (Summer):24–50.

| 2 |

SERIOUSLY VISIBLE

Anne Frances Wysocki

It is old and not uncriticized news that hypertextual documents are by their very structure supposed to encourage readers into more active and engaged relationships with texts and thus with each other. It is also old and not uncriticized news that documents that give more weight to their visual rather than their verbal components ought not to be taken seriously or ought to be relegated to children and the illiterate.

This writing wishes to join the critics of these two different positions by offering responsive counterexamples, for although the news is old, it is still very much present and repeated.

I will first lay out some of the claims for hypertext, and then some of the claims against visual texts, and then provide as counterexamples readings of two pieces of interactive computer-based multimedia.

In a book on hypertext first published in 1992 and then revised and reprinted in 1997, George Landow argues that hypertext:

provides an infinitely recenterable system whose provisional point of focus depends upon the reader, who becomes a truly active reader in yet another sense. One of the fundamental characteristics of hypertext is that it is composed of bodies of linked texts that have no primary axis of organization. . . . Although this absence of a center can create problems for the reader and the writer, it also means that anyone who uses hypertext makes his or her own interests the de facto organizing principle (or center) for the investigation at the moment. (36–37)

Following that description, Landow draws on examples of hypertexts that include fiction, poetry, nonfiction explorations of particular literary works, and pieces meant for entertainment.

Jay David Bolter, in an article published in 1992 but based on a conference presentation in 1989, also speaks of how hypertext changes the relation between author and writer:

In electronic writing, reader and author share in the act of making the text and therefore in the responsibility for the result. . . . In any hypertext, the text originates in an interaction that neither the author nor reader can completely predict or control. (31)

In 1988, Edward Barrett, using cognitive science as his starting point for observation, wrote these notes after observing people at work in an electronic writing classroom:

the traditional classroom unglued—new image for functions of teaching. on-line system as a sequence of interruptions, interaction speeded up, puts you inside the process. from critic to collaborator. Display shows mind in progress. Changing the time scale to get closer to the creative act in composition. (xviii)

More recently, in a chapter in a book published in 1996, J. Randal Woodland (1996) describes how writing on the computer changes the relation between reader and author:

Even in the most rigorously structured electronic books, the very nature of digital text instigates a shift in the locus of textual control from author to reader. (183)

Woodland gives examples from commercial multimedia pieces, some of which are meant to be entertainment, some educational.

To step into another field, here are words from Jakob Nielsen (1990), a computer scientist:

The real issue here is the extent to which the user is allowed to determine the activities of the system. (10)

Certain structures of writing on the computer, then, no matter whether the writing is meant to be educational, literary, entertainment, or functional, give responsibility—or, at least, some or more responsibility—to readers.

But this is not simply an issue of more active engagement with what is on a page. I point you back to the quotation from Barrett above and its implications for classrooms, and I quote Landow (1997) again, who argues that, because of how it shifts responsibility for textual meaning-making, "hypertext answers teachers' sincere prayers for active, independent-minded students who take more responsibility for their education and are not afraid to challenge and disagree" (268). The implication here is that the heightened responsibility of constructing meaning out of a text, of not having meaning (the meaning) of a text handed to one by the writer, will encourage readers—students—to take more active roles in their education, to question not just texts, but everything. The implication is that readers—students—through having to grapple with meaning making, will develop stronger senses of their varied and particular positions and possibilities and hence will not acquiesce unquestioningly to other positions.

The relation that Landow implies here—between having and holding strong opinions and then desiring and making strong active learning—I see echoed in other writers concerned with political structures outside the classroom. I will come back to consider the force of several of the following arguments in more detail later, but for now I want only to invoke them here, first by stretching back to John Stuart Mill (1975): in *On Liberty*, Mill argues that "Liberty of Thought"—by which he means that "human beings should be free to form opinions, and to express their opinions without reserve" (53)—is "part of the political morality of all countries which profess religious toleration and free institutions" (16):

> Who can compute what the world loses in the multitude of promising intellects combined with timid characters, who dare not follow out any bold, vigorous, independent train of thought, lest it land them in something which would admit of being considered irreligious or immoral? (33)

More recently, Jürgen Habermas (1989) argues in *The Structural Transformation of the Public Sphere* that the "public sphere of a rational-critical debate" he believes to have existed in eighteenth century England depends on a particular kind of subjectivity; that subjectivity developed, for Habermas, through the close interpretative work of individuals reading novels, by which each "attained clarity with itself" and so was able to have the personal opinion necessary for (for Habermas) real public debate (43–56). Paul Virilio argues in both *The Art of the Motor* (1995) and *The Vision Machine* (1994) that human political freedom depends on our not all sharing the same opinions; such freedom depends then, for Virilio, on our not accepting the repeated opinions that are given to us through the communication media surrounding us. Andrew Feenberg argues in both *Critical Theory of Technology* (1991) and *Alternative Modernity* (1995) that modern technopolitical structures push us toward uniformity of opinion, but that "new degrees of freedom" are possible only with "multiple standpoints and aspirations" (1995, 231–232). Especially the last three of these writers would each take issue with the particular routes the others take in their arguments, but I want to emphasize that—broadly—each believes that a diversity of strong positions and viewpoints is necessary for any kind of free political realm; each believes that when—from the causes described—we hold standardized opinions and thoughts, then commercial or political entities are more able to build, unopposed, the worlds that are efficacious for them but stifling of diverse human ability.

And I want to emphasize that I believe the weight of the claims I have quoted so far about how we read electronic writing lies in their connections to such politically charged arguments as I have just outlined: The skills and opinions a reader of such texts learns in the classroom (if the claims I have quoted for the effects of such reading are accurate) are those skills and opinions central to the kind of "political morality" I believe many of us desire. Because electronic texts are to give their readers responsibility for constructing meaning, those readers are to be supported in forming the strong and independent opinions that help them sustain the political freedom underlying a democratic society. The goals that I believe to underlie the hopes for hypertexts thus articulate classroom work to political stance, implying that the practices people acquire in our classrooms have direct effects on their contributions to our civil society.

I am aware of the quickness of my preceding arguments, and of the many places where you—if you are being the active reader the preceding paragraphs describe—will have raised objections to my quicknesses or to my jumping blithely over the openings for objections, for almost as soon as the above hopes for electronic writing were being formulated, objections to those hopes—tied to the promises of those hopes—were appearing. I am well aware, for example, of arguments like those of Martin Rosenberg (1994), that computers cannot support the kinds of freedoms I have been describing others describing because computers are themselves the products of the larger sociopolitical forces that work against such freedoms. I am well aware of Myron Tuman's (1992) critique of the belief that "technology itself, like miraculous new medical equipment for seeing inside the human body, will solve our problems—make us a healthier society—without having to be embodied in institutions of social reform" (133). I am well aware of Espen Aarseth's (1997) similar critique that not only is electronic writing alone not sufficient for achieving the hope I have described, but there is nothing to stop paper texts from being shaped toward supporting a reader's freedom and electronic texts toward opposite ends:

Electronic writing can just as easily (probably even more easily) maintain the two-level hierarchy of a flat expression plane, whereas preelectronic writing (e.g., Queneau's *Poèmes*) [can] subvert it. (176)

I am well aware that Landow (1997) himself has modified his original hopes (14–33) to argue that only certain kinds of computer writing—that which is networked and allows readers to add their own comments to other texts—support the hope I have shown others to express. (For a rich and critical expansion of many of these concerns, as well as of others that are not as pertinent to my particular project, please see Johnson-Eilola's [1997] *Nostalgic Angels*.) I am, finally, more than aware that the last three writers whose arguments I cited above on the necessity of multiple standpoints for political freedom all see technology—especially communication technologies such as computers and the news and entertainment media—as seriously implicated in a diminishment of possible standpoints.

What these criticisms address is a certain hermeticism in the hopes I have quoted for hypertexts, an implicit belief that exposure to hypertexts alone—that simply reading hypertexts—will automatically give readers the qualities of active independence. What the criticisms have in common is that they all ask us to be mindful that no technology is autonomous and that we use no technology outside the webs of its history, of its connections to other technologies, or of our motley and complex relations with others.

I'll be returning to the issues about hypertexts, but I wish now to develop the other stream of argument—concerning the visual—that addresses the kinds of texts I'll be considering shortly.

A few years ago at a panel on assessing electronic writing at the annual Conference on College Composition and Communication, an audience member asked the panelists, "When we teach students to make multimedia, aren't we just teaching them to make Levi's 501 ads?"

The concern behind this question is also present in the words of Habermas (1989) when he argues in *The Structural Transformation of the Public Sphere* that

The techniques of the cartoon, news pictures, and human-interest story grew out of the repertory of the weekly press, which even earlier had presented its news and fictional stories in a way that was as optically effective as it was undemanding at the literary level. . . . By means of variegated type and layout and ample illustration reading is made easy at the same time that its field of spontaneity in general is restricted by serving up the material as a ready-made convenience, patterned and predigested. (168–169)

Habermas, in his consideration of what makes possible an active and effective public sphere, believes that only individuals who have worked hard in the act of reading words-only literary books will have acquired a sufficiently strong sense of self and opinion to make the thoughtful intellectual contributions necessary for the public sphere he desires, as I have described above. For Habermas, then, when words are given visual inflection or are replaced by pictures, as in cartoons, the required intellectual work is no longer required, and people who read such
materials
are rendered
passive
and
weak.

Paul Virilio (1994) makes similar observations about our present technologies of communication: According to Virilio, when we are all busy looking through glass—looking into camera lenses and shop windows or television sets as through windows onto other lives—our technologies of communication create

not only window-apartments and houses, but window-towns and window-nations, media megacities that have the paradoxical power of bringing individuals together long-distance, around standardized opinions and behaviors. (65)

Virilio is arguing, in the text from which the quotation comes, that because any time for reflection has been taken away from us by the speed of communication technologies, we are pliable before ready-made opinions; the various different possible re-presentations (as Virilio often puts it) we might make by reflecting on our sights are cut off before they are begun, replaced by images that do not (according to Virilio) need reflection and re-presentation: “geometric brand images, initials, Hitler’s swastikas, Charlie Chaplin’s silhouette, Magritte’s blue bird, or the red lips of Marilyn Monroe” (14).

As in response to the claims that hypertext automatically make active readers, there are criticisms in response to these claims about the automatically enervating visual. W. J. T. Mitchell (1986), James Elkins (1999), and Wendy Steiner (1982), for example, have variously explicated and critiqued the histories and cultures that lead us to believe words are serious and images not; their arguments are similar to those concerning hypertexts that remind us that we ought not to consider technologies outside their social and temporal contexts. Barbara Maria Stafford (1996) and Gunther Kress and Theo van Leeuwen (1996), meanwhile, have argued for how much cultural and intellectual weight image-heavy texts can carry. Hugh Kenner (1994) discusses the differences between the animations of Warner Brothers and those of Disney to show how they encourage very different responses to our worlds, while John Berger (1980) compares (bleakly) Jiminy Cricket to Francis Bacon's paintings; see also the widely ranging work of the Society for Animation Studies, in the collection edited by Pilling (1997).

Rather than dig into these critiques, however, I want to emphasize how the two streams of argument I have described weave into each other thickly. The concerns in the arguments for hypertext and against the visual entwine in their concerns over the necessity of interpretation for active, engaged citizens: In these arguments, hypertexts necessarily require interpretation and visual texts do not. When we rely on the visual for communication, Habermas (1989) and Virilio (1994) argue, whether we see the visual elements on paper or through some kind of screen, everyone knows the same things: Texts that use visual arrangements to support their arguments predigest and standardize those arguments, removing from them any possibility of differing interpretation. They make those arguments too easy by not asking of us the struggles of interpretation that help us develop the differing positions that Habermas and Virilio, as I have earlier cited them alongside Mill and Feenberg, believe necessary for the functioning of democratic societies. In all of these arguments, then, starting with the argument about multimedia's being like Levi's 501 ads, the concern is that the visual is not serious enough, that its particular and enfeebling topics are advertising and cartoons and propaganda and sex. The assumption behind the critique of the visual is that we each take in what we see, automatically and immediately, in the exact same way as everyone else, so that the visual requires no interpretation and in fact functions as though we have no power before it (see Rutledge's [1994] description, for example, of how advertising functions: Students are ducks, targets for advertiser-hunters); the assumption behind the celebrations of hypertext is that any text that presents us with choice of movement through it necessarily requires interpretation.

I am not going to add to the critiques of these positions by taking up the important perspective the critiques I have listed call us to remember: that we cannot separate our technologies, including our technologies of communication, from their contexts. Instead, I wish to add to these critiques by doing close rhetorical readings of two multimedia texts, *Scrutiny in the Great Round* (Dixon, Gasperini, and Morrow 1995) and *Throwing Apples at the Sun* (Earls 1995), in order to show, perhaps contradictorily, that these texts require focused interpretation of the visual at the same time they work to mitigate any hope that interpretation necessarily leads us to active engagement with ideas and each other. I am not presenting either piece of multimedia as a representative of all multimedia, but I am considering them here because they are examples of serious argumentative compositions whose makers are attentive to the visual possibilities of the technologies they use but who argue against the possibilities and efficacy of liberal political engagement tied to interpretation.

Scrutiny in the Great Round was published in 1995 by Calliope, with authoring credit given to three people, Tennessee Rice Dixon, a maker of artists's books, Jim Gasperini, a programmer and software designer, and Charlie Morrow, an audio designer. Mark Bernstein (1998) of Eastgate Systems has called *Scrutiny in the Great Round* "a vast dynamic painting, constantly changing and responding to the viewer," and it is to the aesthetic attentions of this piece that I will return, after stepping you through the piece.

On the computer screen, when I double-click the icon for *Scrutiny in the Great Round* on the desktop, the whole screen is taken over, turned to black. There is no longer a menu, and there is thus no way for me to run other applications at the same time. I am either in *Scrutiny in the Great Round* or completely out of it: There are no halfway conditions here.

Out of the blackness comes a short opening animation in which the publishing company's name dances a bit on screen, and then there is a screen with a star-sieved night sky on it, and over the stars is layered a sketchy, lightly-tinted drawing of several leaf-less trees, symmetrically balanced; there are also two round shapes almost symmetrically balanced near the top of the screen, and there is the title in the middle (figure 2.1).

It turns out that this is not a title screen that functions as title screens do in most multimedia, as a visual introduction to be left behind: It turns out that I am instead on a screen that functions like all the other screens of this piece; I am on a screen that is just "one more" of the twenty-three screens that make this piece. It turns out, I discover through further exploration, that the two round shapes on this screen are zygotes, and that I am in the middle of what Bernstein (1998) calls "a grand cycle of collages, each representing a moment in the progress of regeneration: encounter, romance, gestation, emergence."

If I move the cursor over this screen, it changes to indicate that different parts of the screen are clickable, but it also changes into different objects to indicate that different things will happen if I click. At the top of the screen, the cursor changes to something that looks a bit like a patchwork quilt: clicking the cursor here beings me to a

| **Figure 2.1** |

Opening screen of *Scrutiny in the Great Round.*

"map" (figure 2.2) for the "grand cycle" Bernstein names, on which I can see how the screens of the piece are arranged into two concentric circles/cycles, and on which I can click any of the smaller screen representations to go to a particular screen.

If, back on the opening screen, I move the cursor to the bottom of the screen, it becomes a long, braided chain—DNA shaped—and clicking there steps me through all the screens of both cycles, automatically playing all the piece's animations for me. If I move the cursor to the left or right, it changes to a left- or right-facing bird, and if I click when the cursor is a bird, I move to the screen that precedes or follows this one in the "grand cycle." If I move the cursor over the zygotes, the cursor changes to a sun (left zygote) or moon (right zygote): Clicking the cursor when it is the sun results in an animation that plays over this screen; clicking the cursor when it is the moon takes me to an "inner," moon screen of the concentric circle of the map, to the screen that

| **Figure 2.2** |

"Map" screen of *Scrutiny in the Great Round.*

corresponds to this "sun" screen. (On such inner/dark/moon screens, moving the cursor left or right gives the same results—except that the left- and right-facing bird cursor is replaced by a fish cursor.) As some of my words here have implied, it turns out that the two concentric circles of the cycle are "about" higher and lower stages in the cycle of life.

I'm going to write now about one particular of the twenty-three screens of this piece—the "Nesting" screen, in between the "Pregnancy" and "Parturition" screens—to give you a sense of how all the screens in the cycle function. (I want to point out here, before I start my description, that the visual effect the authors of this piece chose to use as I move between the screens of this piece is almost always a dissolve, as though what is on one screen becomes what is on the next, slowly: It is as though I stay in the same place and watch images and processes transmute into each other. I could say that one argument of this visual transition, then, is that everything is made of the same, meta-

| **Figure 2.3** |

Sun "Nesting" screen of *Scrutiny in the Great Round.*

morphosing material.) The sun "Nesting" screen (figure 2.3), the one I wish to discuss as representative, shows a lightly yellow-tinted rectangle against the same starry night sky that is the background for all the screens: This background gives the visual sense that the night sky extends out infinitely behind all the screens, dwarfing what is there, as though we have been focused on one small process out of many possible others.

This tinted rectangle is veined with smudges, and on it appear biological drawings of mitosis and cell separation, a horse, a bird, a fish, some fruit, some plants, a vase, a circle. The circle is filled with images that are hard to discern but that look like old wood cuts, but there is in the circle one unmistakable image: that of an egg, with some linked rings over it.

All these images appear variously on other screens; it doesn't take long to figure out as I move through the different screens that the horse represents the masculine, and circles and vases the feminine, with the fruit being the results of their union. There is nothing new in this.

And I do not think the authors intended there to be anything new in this.

The different images I have mentioned as being on the different screens are not separated or made to appear as though they have some kind of visual or spatial depth between them: they have been worked into a flattened collage in which everything touches and blends and dissolves into everything else. These images come from a wide variety of sources: old European woodcuts and nineteenth-century paintings, Mayan and Incan and Greek sculptures, old botanical and biological drawings. All is grist for the mill here, everything repeats with slight variations, and screens echo each other visually, suggesting that all the stages carry something of all the other stages within them.

If, back on the sun "Nesting" screen, I click the egg, a softly circular cutout of a naked woman, embraced by a statue of a Greek goddess (a constructed image) comes onto the screen from top right and moves to cover the egg. Then, through a slow and teasingly beautiful animation—in which the circular shape shifts into a spiral—the circle shifts into a fetus, which itself then spirals into and dissolves slowly and suggestively back into the egg. Many of the animations of *Scrutiny in the Great Round* involve such spiraling, a circling in or out: Sea horse tails become spiral staircases, horses spiral into men—all repeating the overall, inescapable circularity of the multimedia piece itself.

I want to move out from *Scrutiny in the Great Round* now, to escape it in order to speak about it, from outside. This could be an oppressive piece of multimedia: There is the visual strategy of taking over the whole screen so that it is as though the piece becomes my whole world at the time of my interaction, there is the visual darkness of the piece, there are the inescapable circularity and repetition in design and interaction, and there are then the implications that the circularity is all there is, the dissolving of one thing into another all that there is. The implication is that romance is for biological purposes only and exists only as the generative midstep between death and birth, no matter where or when. This is the stuff of Schopenhauer.

And the authors of this piece do take a Schopenhauerian attitude toward what they present, through the possible interactions they give me: there is for them, for me, no way out, except for aesthetic appreciation.

As Schopenhauer (1969) argued, carrying Darwin's structures of interpretation to a logical conclusion, there is nothing in the world but the imperative to reproduce. Under this reading, my actions might seem self-directed and independent, but only to me. In the big scheme of things, all that I do, according to Schopenhauer's carrying out of Darwin, serves simply to keep the biological world ticking and revolving along. There is no breaking out of the cycle. Whatever is produced serves the re-producing design, whatever is consumed is re-cycled. The logical and emotional position to take in the face of this, Schopenhauer argued, is aesthetic resignation: We are to appreciate the beauty and to accept that we can neither deny nor challenge our lack of independence. Instead, argues Schopenhauer, we should

turn our glance from our own needy and perplexed nature to those who have overcome the world, in whom the will, having reached complete self-knowledge, has found itself again in everything, and then freely denied itself, and who then merely wait to see the last trace of the will vanish with the body that is animated by that trace. Then, instead of the restless pressure and effort; instead of the constant transition from desire to apprehension and from joy to sorrow; instead of the never-satisfied and never-dying hope that constitutes the life-dream of the man who wills, we see that peace that is higher than all reason, that ocean-like calmness of the spirit, that deep tranquillity, that unshakable confidence and serenity, whose mere reflection in the countenance, as depicted by Raphael and Correggio, is a complete and certain gospel. Only knowledge remains; the will has vanished. (411)

And so I can offer this interpretation of *Scrutiny in the Great Round*, based on my attentions to its visual pieces and structures and interactivity: All the visual strategies of this piece work against suggesting any other possibility for coming to terms with the root facts of life and death other than an aesthetic resignation, as Schopenhauer describes. In *Scrutiny in the Great Round*, I can move from screen to screen, taking delight in the ingenious and soothing and beautiful and slow animations, taking pleasure in the intellectual and visual connections I can make between the different images and animations. As it is for Schopenhauer, the kinds of interactions I as audience member am allowed in this piece—the ones that soothingly distract me from the rigid and inescapable physical necessities of life—are the ones that encourage my aesthetic contemplation of what is around me.

Before I speak more specifically of how *Scrutiny in the Great Round* serves as counterexample to the two streams of argument I earlier set out, let me now turn to the second piece of multimedia, to *Throwing Apples at the Sun*, a piece that provides an edgy Lyotardian, postmodern counterpoint to *Scrutiny in the Great Round*, *nomos* to *physis*, a much "messier" piece visually. *Throwing Apples at the Sun* is, like *Scrutiny in the Great Round*, copyrighted 1995, but is authored by one person, Elliot Earls, a graduate of the Cranbook Academy with an M.F.A. in graphic design. The title of this second CD—the suggested attempt to touch the sun with a piece of fruit, and specifically with the

---

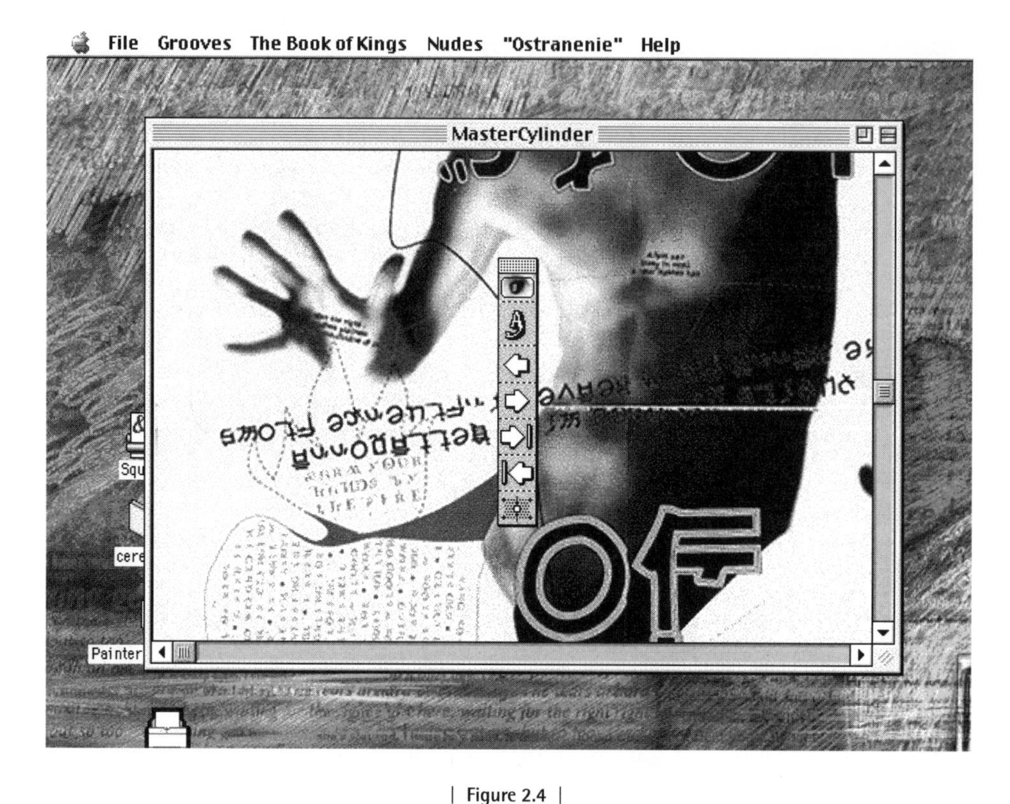

| Figure 2.4 |

"Master Cylinder" screen of *Throwing Apples at the Sun* (showing the desktop of my computer behind the windows of the multimedia piece).

representative piece of fruit in Christian mythology, an apple—might lead me to expect an attitude of similar futility toward human endeavor as that depicted in *Scrutiny in the Great Round*, and there is futility, but it is different in color and shape. It is cultural, rather than "natural" as it is in *Scrutiny in the Great Round.*

From the first, *Throwing Apples at the Sun* is presented and structured very differently from *Scrutiny in the Great Round.* After a quick title screen, I see a window named "Master Cylinder" (figure 2.4) and a small palette window with some apparently standard navigational arrows on it. This piece does not take over my screen as does *Scrutiny in the Great Round;* instead, I can see my computer desktop behind the two windows that open, and there is a menu bar at the top of the screen with (except for the first two items) items that are in no other Macintosh menu I know: After the usual "Apple" and "File" choices, there are "Grooves," "The Book of Kings," "Nudes," and "'Ostranenie'." And what's listed under the menus are not the usual options, either: Under "Nudes," for example, the options are "5 Nudes," "A Dead President," and "Shell of a Man." I have many options here, it seems, and no directions. Things are (apparently) under my control.

If I click the right-facing arrow in the small palette (which, like the other window, I can move wherever I want on screen), what is in the "Master Cylinder" window changes. The window is not large enough for me to see all that's in it: I must scroll to see what's there, or I can make the window larger if I want (and if the size of my monitor allows for this). What is in the "Master Cylinder" window are images of a set of posters that come with the *Throwing Apples at the Sun* CD-ROM, but unlike the paper versions, these images are clickable and may yield small video clips (which may take place in separate windows or within the "Master Cylinder" window itself), or they may yield dialog boxes, but these are dialog boxes that don't carry the usual kinds of messages like "Are you sure you want to quit?" Instead, a *Throwing Apples at the Sun* dialog might ask, "The logical as path way to dysfunction . . . ?" (figure 2.5), with my only possible response being "OK."

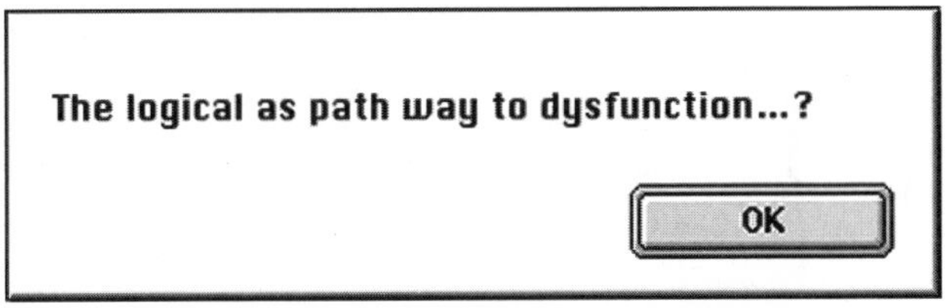

The logical as path way to dysfunction...?

OK

| **Figure 2.5** |

Dialogue Box.

That "dys" prefix shows up a lot in *Throwing Apples at the Sun*, in "dysplasia" and "dyslexia" and "dyspepsia," words that are peppered throughout the piece and that are also the names of the typefaces Earls made to use with this piece (and that comes included with the CD).

Earls has used these typefaces in various ways in this multimedia piece. For example, if I choose "5 Nudes" from the "Nudes" menu, a new window, titled "5 Nudes and a Dead President" (figure 2.6) (although there are only four nudes) opens over the other windows, showing male nudes in awkward positions, imagery suggestive of the series of paintings Robert Longo made in the 1980s of his friends responding to having things thrown at them or—to reach further back—suggestive of the many paintings of Saint Sebastian crumpled around the arrows in his body. (This crumpled male body imagery is repeated in the poster in the "Master Cylinder" window and in various other of the small windows and video clips that appear in the piece.) These nudes in the "5 Nudes and a Dead President" window have been "censored," with boxes over their genitalia and faces. Those censoring boxes are used elsewhere in the piece, as I will soon discuss, but here they are generative because they are clickable: My clicks do not remove the boxes, but instead bring up short phrases, such as "On the road to Damascus" or "The

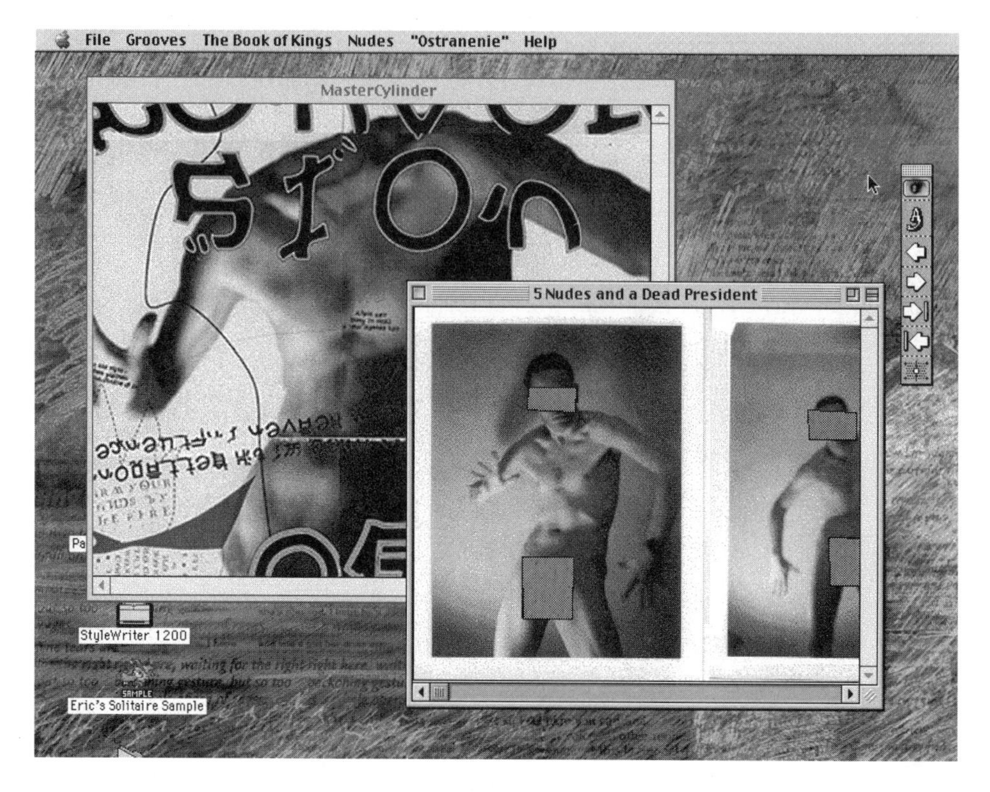

| Figure 2.6 |

"Five Nudes and a Dead President" screen of *Throwing Apples at the Sun.*

Heretic is blind" or "The Heretic is dumb," and the typefaces of these phrases are awkwardly shaped, dyspeptic themselves. Their appearance comes to me as though I had touched a concentrating person on the shoulder, and he had jumped in response to my touch.

The typefaces, in their words, also call, obviously, on the experiences of Saul as he is transformed into Paul . . . but in this window, if we scroll to the end of the "path," we don't have the usual sainthood: Instead, there is a painting of John F. Kennedy, who, when I click him, is surrounded by a red and black scratchy halo and, at bottom, a red blotch, as of blood.

If I continue to explore this piece, desiring, as I do, to see all the windows and clickable things—feeling that if I could only get it all out on screen at once I could have it all, understand it all—if I continue, eventually I will come to the "Book of Kings" win-

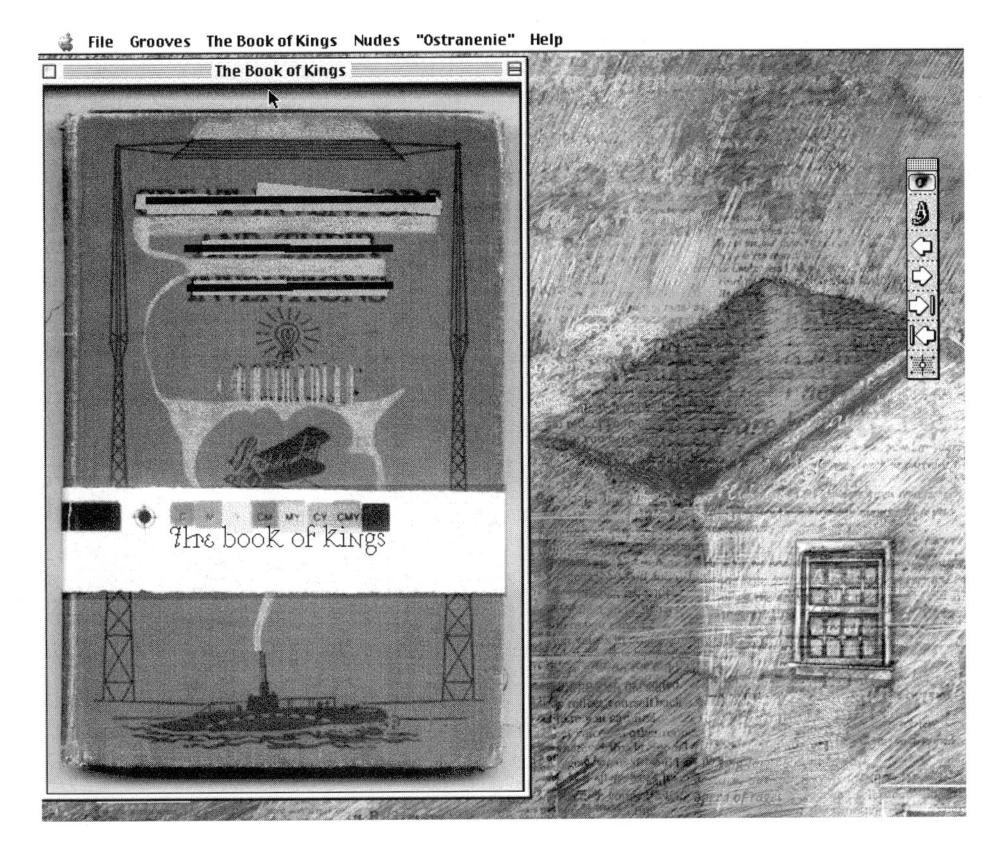

| Figure 2.7 |

"Books of Kings" screen of *Throwing Apples at the Sun.*

dow (figure 2.7). I can get to this window (as I can get to most parts of *Throwing Apples at the Sun*) by a menu choice or by a click on something in another window. There are no dissolves between sections or animations as there are in *Scrutiny in the Great Round;* there are only jumps in this piece, with windows appearing or disappearing, text appearing or disappearing, windows abruptly resizing before my eyes. The "Book of Kings" is a window that shows the cover and various pages of a book titled *Great Inventors and Their Inventions,* except that the title and various phrases of the book (as with the earlier genitalia and faces) have been covered with censoring black boxes and are so hidden from me—but incompletely. As with the genitalia-covering boxes, I can usually figure out what's under them. The gesture to hide is futile.

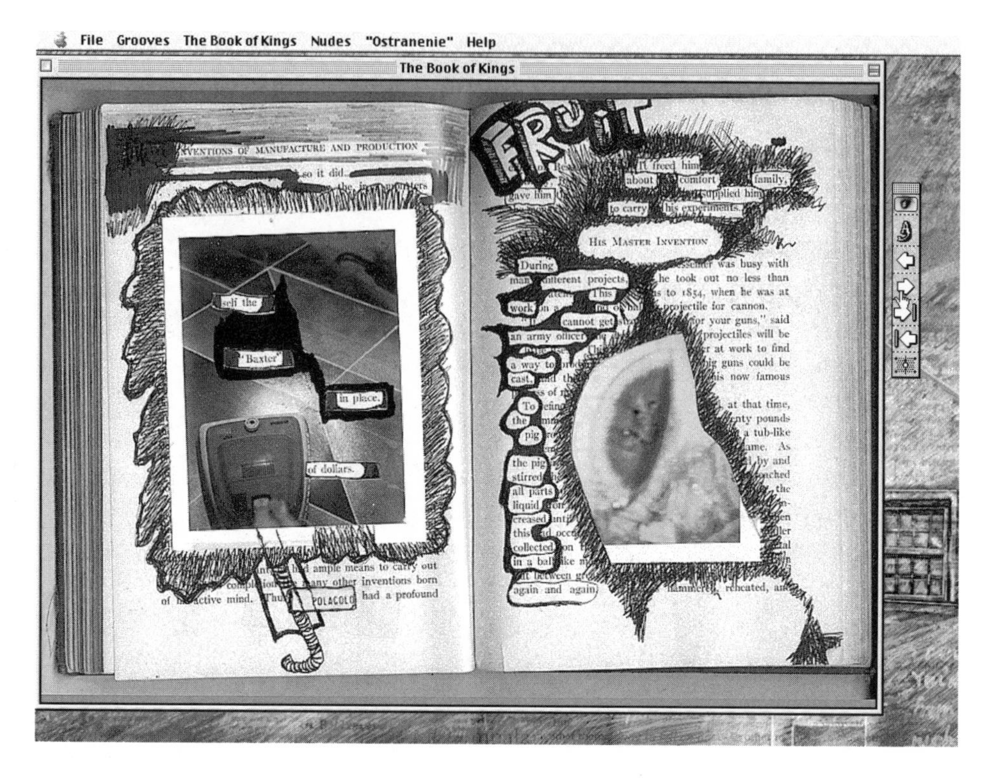

| Figure 2.8 |

Moving into the "Book of Kings" window of *Throwing Apples at the Sun.*

The book about inventors (figure 2.8) is from the 1920s, full of optimistic stories of "great inventions" and "famous inventors," but Earls has drawn over these pages, counterpoised the notions of greatness with (for example) images of vacuum cleaners on messy tile floors. And here, as in other windows and the title of the piece the notion of fruit appears, just as in *Scrutiny in the Great Round.* But here, rather than suggesting the beauty of reproduction, fruit is a word, scribbled or splattered, used in phrases like "the fruit of our labors, the labor of pain" and linked to vacuum cleaners, not babies.

I hope I have played *Scrutiny in the Great Round* off against *Throwing Apples at the Sun* enough so that I can argue here that, if the visual strategies of *Scrutiny in the Great Round* are about finding distraction and individual solace in aesthetic reflection, *Throwing Apples at the Sun* seems to be about the futility but final necessity of making things to reflect on. If nature leads to death but eventual rebirth in the first piece, culture in the second piece leads to Bissell and to St. Paul by the grassy knoll in Dallas, but there is the inventiveness of *Throwing Apples at the Sun* itself, and the curiosity it encourages. The inventiveness and curiosity are finally proved scratchy, nothing great or all encompassing, but still worthy of pleasurable engagement (at least for some). Matthew Kirschenbaum (1998), for example, ends his consideration of this piece—a consideration that places *Throwing Apples at the Sun* in the context of other works that foreground the material conditions of their making—with these words:

> Because of the opacity of the interface . . . it is often unclear whether a given visual or audio event—the appearance of a window, a jump in the sound track—is the result of some action on the part of the user, or a process running deep within the program's logic. The effect is uncanny, but also surprisingly tranquil: rather than becoming frustrated at my attempts to manipulate the work, I find my interactions deeply soothing, the equivalent of a human-computer multimedia jam session. (sec. IV)

If *Scrutiny in the Great Round* offers reflection and slow and beautiful dissolves between scenes and the opportunity to lose oneself in the reflection, *Throwing Apples at the Sun* offers edgy complexity and distraction through wondering—through asking me to wonder—what might pop up next when I click something. What other connections can be made, how else can hand and mouse be put to work (if always dyspeptically, futilely)?

I can't get the entire piece out on screen at once with *Throwing Apples at the Sun* as I can with the "map" in *Scrutiny in the Great Round.* I can only muddle through, stumbling awkwardly—or soothingly, as Kirschenbaum (1998) reads the piece—through ungainly typefaces and jumbled-up windows. There is no way to encompass everything all at once in some great understanding, as with my Schopenhauerian take on *Scrutiny in the Great Round.* Instead, I get piecemeal views, bits of things that suggest possibilities of new connections, which are always broken by being in different windows or scratched over. The interactions of this piece, as with the first, work to deny—if through different strategies and by creating a very different mood—that there is much if any hope for active, independent human engagement with the world.

Now let me speak specifically (if I have not been suggestive enough) of how I think these two pieces of multimedia serve as counterexamples to the two streams of argument I laid out back at the beginning of this writing.

I'll respond first to the stream of argument that comes out of observations about hypertexts. The claim, you'll remember, is that the design of hypertexts and multimedia—

the requirement of the interactive structure that readers must choose their own paths through the work—would necessarily produce active, challenging readers because those readers would each come to their individual interpretations of a text by the design's requiring them to choose individual navigational paths, hence individual paths of understanding. But what then to make of such pieces as I just showed you, where the interactivity is designed to persuade readers that being active and challenging is futile?

The choices the two pieces of multimedia offer their readers for moving through the texts work such that any path a reader takes is an argument against seeing or finding any worth in active, independent engagement with the world and each other; the visual structures of these pieces argue instead, through their shaping of the choices and structures they offer readers, for taking pleasure in, almost, soliloquy, in passive, soothing, personal inward enjoyment of the different aesthetic and cultural-object combinations the world has to offer.

The design of these pieces, I am arguing, is a direct counterexample to the arguments I laid out earlier for hypertext and multimedia.

But in order to make the claims that I have made to counter the arguments for hypertexts, I have had to interpret the pieces I showed you. I didn't present them to you without commentary, I didn't come to my understandings of them through simply seeing them and having an immediate comprehension of some simple, predigested cultural construct. When working through these multimedia pieces I had to apply abilities and approaches and understanding from rhetoric and art and visual communication and philosophy and critical theory; I had to bring to them understandings I've acquired through working with other electronic texts. I presented my understandings of these texts to you in (what I hope has been) clearly and coherently structured arguments—but behind those arguments lies considerable time and messy effort in interpreting how the parts of those texts and my movements through them combine into some (perhaps temporary) coherent opinion of them. And I am certainly not arguing that my readings of these pieces are the only possible ones or that, were I to spend different time with them elsewhere, I'd come to exactly the same readings.

But all of that—what I brought to these pieces in order to read them and the way that I read them—argues against (I hope) the arguments of Habermas (1989) and Virilio (1994) that visual texts make reading easy, that they predigest meaning, or that they make for standardized opinions and behaviors. I hope that my interpretations have persuaded you that, at least sometimes, visual texts can be as pleasurably challenging as some word-full texts.

In closing, I acknowledge that (as I indicated earlier might happen) I may seem to have worked myself into some kind of contradiction: I am claiming here that interpreting these pieces requires considerable effort to build the interpretation and that the pieces' visual structure and design finally discourage readers from believing that such effort is efficacious. Rather than contradiction, though, I see this as a tension here, tension between means and ends, between the systems and structures and strategies and work of argument and the position to which that argument hopes to persuade a reader. It is a tension that points away from the expansive and strongly stark streams of argument I laid out at the beginning of this chapter—the arguments that imply that visuals and hypertexts and multimedia must always accomplish the exact same things everywhere—and it is a tension that points toward the very different kinds of things we can accomplish with the materials (words and images and visual arrangements) with which we can compose.

But we can compose in new ways only if we acknowledge that the visual and hypertextual aspects of our texts are not monolithic. Even to say "the visual" or "the hypertextual" is to imply that anything that fits under one of those signifiers points to the same signified; the pieces of multimedia I have analyzed show this not to be the case. Yes, there are visual objects that seem to be readily identifiable and understandable anywhere, but they have been designed to be that way (think "Coca-Cola," with 80 percent of its revenues coming from outside the United States and using a very aware set of strategies to make the logo be recognized "everywhere" [*The Coca-Cola Company* 2000]); they have been purposefully and awarely fit into larger cultural and global technological and communication structures that enhance their "instant" recognition and standardization. If we want our texts to be complex and to ask for interpretation, there is nothing inherent in "the visual" or "the hypertextual" demanding this or standing in our way—except beliefs in some inherent simplicity of "the visual" or complexity of "the hypertextual." If we want our students to value active engagement with texts and each other, we cannot expect that our texts will do that in and of themselves. If we find our students making Levi's 501 ads when we ask them to make multimedia, it is not the technologies of multimedia determining the outcome, but rather, in no small part, what we have taught them about the potential complexities and contexts of texts that incorporate multiple media. If we want there to be more complex texts in the world and more complex and active readers and citizens, then let's work with people in our classes to make such texts and to develop together the abilities and concerns to help us be the latter.

Acknowledgments

I presented earlier versions of this paper at the Modern Language Convention, December 1998, San Francisco, at the University of Waterloo, Ontario, Canada, and at the University of Illinois at Urbana-Champaign. I thank all the thoughtful listeners and questioners who helped me sharpen the directions of this writing.

Works Cited

Aarseth, Espen J. 1997. *Cybertext: Perspectives on Ergodic Literature.* Baltimore: Johns Hopkins University Press.

Barrett, Edward. 1988. "Introduction: A New Paradigm for Writing with and for the Computer." In *Text, Context, and Hypertext: Writing with and for the Computer,* ed. Edward Barrett, xiii–xxv. Cambridge: MIT Press.

Berger, John. 1980. "Francis Bacon and Walt Disney." In *About Looking,* 111–118. New York: Pantheon

Bernstein, Mark. 1998. "The Hypertext Patterns of Scrutiny in the Great Round." Web site. Available: <http://www.eastgate.com/HyperTextNow/archives/Scrutiny.html> (accessed December 15, 1998).

Bolter, Jay David. 1992. "Literature in the Electronic Writing Space." In *Literacy Online: The Promise (and Perils) of Reading and Writing with Computers,* ed. Myron C. Tuman, 19–42. Pittsburgh: University of Pittsburgh Press.

The Coca-Cola Company: The Real Global Thing!" 2000. Web site. Available: <http://www.lippincott-margulies.com/sense/s95_coke1.html> (accessed August 25, 2000).

Dixon, Tennessee Rice, Jim Gasperini, and Charlie Morrow. 1995. *Scrutiny in the Great Round.* CD-ROM. Santa Monica, CA: Calliope Media.

Earls, Elliot. 1995. *Throwing Apples at the Sun.* CD-ROM. Greenwich, CT: The Apollo Project.

Elkins, James. 1999. *The Domain of Images.* Ithaca: Cornell University Press.

Feenberg, Andrew. 1991. *Critical Theory of Technology.* New York: Oxford University Press.

Feenberg, Andrew. 1995. *Alternative Modernity.* Berkeley and Los Angeles: University of California Press.

Habermas, Jürgen. *The Structural Transformation of the Public Sphere: An Inquiry into a Category of Bourgeois Society,* trans. Thomas Burger. Cambridge: MIT Press.

Johnson-Eilola, Johndan. 1997. *Nostalgic Angels: Rearticulating Hypertext Writing.* Norwood, NJ: Ablex.

Kenner, Hugh. 1994. *Chuck Jones: A Flurry of Drawings.* Berkeley and Los Angeles: University of California Press.

Kirschenbaum, Matthew. 1998. "Machine Visions: Towards a Poetics of Artificial Intelligence." In *Electronic Book Review* 6. Available at: <http://www.altx.com/ebr/ebr6/6kirschenbaum/6kirsch.htm> (accessed December 12, 1998).

Kress, Gunther, and Theo van Leeuwen. 1996. *Reading Images: The Grammar of Visual Design.* London: Routledge.

Landow, George P. 1997. *Hypertext 2.0: The Convergence of Contemporary Critical Theory and Technology.* Baltimore: Johns Hopkins University Press.

Mill, John Stuart. 1975. *On Liberty*, ed. David Spitz. New York: W. W. Norton.

Mitchell, W. J. T. 1986. *Iconology: Images, Texts, Ideology.* Chicago: University of Chicago Press.

Nielsen, Jakob. 1990. *Hypertext & Hypermedia.* San Diego, CA: Academic Press.

Pilling, Jane, ed. 1997. *A Reader in Animation Studies.* Sydney: John Libbey.

Rosenberg, Martin. 1994. "Physics and Hypertext: Liberation and Complicity in Art and Pedagogy." In *Hyper/Text/Theory*, ed. George P. Landow, 268–97. Baltimore: Johns Hopkins University Press.

Rutledge, Kay Ellen. 1994. "Analyzing Visual Persuasion: The Art of Duck Hunting." In *Images in Language, Media, and Mind*, ed. Roy F. Fox, 204–208. Urbana, IL: National Council of Teachers of English.

Schopenhauer, Arthur. 1969. *The World as Will and Representation*, vol. 1, trans. E. F. J. Payne. New York: Dover.

Stafford, Barbara Maria. 1996. *Good Looking: Essays on the Virtue of Images.* Cambridge: MIT Press.

Steiner, Wendy. 1982. *The Colors of Rhetoric: Problems in the Relation between Modern Literature and Painting.* Chicago: University of Chicago Press.

Tuman, Myron. 1992. *Word Perfect: Literacy in the Computer Age.* Pittsburgh: University of Pittsburgh Press.

Virilio, Paul. 1994. *The Vision Machine.* Bloomington: Indiana University Press.

Virilio, Paul. 1995. *The Art of the Motor.* Minneapolis: University of Minnesota Press.

Woodland, J. Randal. 1996. "Spider Webs, Symphonies, and the Yellow-Brick Road: Form and Structure in Electronic Texts." In *The New Writing Environment: Writers at Work in a World of Technology*, ed. Mike Sharples and Thea van der Geest, 183–203. London: Springer-Verlag.

| 3 |

THE DIALOGICS OF NEW MEDIA: VIDEO, VISUALIZATION, AND NARRATIVE IN RED PLANET: SCIENTIFIC AND CULTURAL ENCOUNTERS WITH MARS

Helen Burgess, Jeanne Hamming, and Robert Markley

In 1997, Robert Markley was invited by a prominent colleague at another university to contribute a module about Mars to a proposed CD-ROM history of science fiction. She had received funding to develop a pilot version of this project as an on-line course: Students would pay a fee to the school for the CD-ROM or for access to its contents over the Web. The course would consist of fifteen modules, each of which would feature video interviews with one or two prominent science fiction authors, and focus on three key works. Having just begun a book about Mars in science and science fiction, Markley agreed, secured start-up funding, and began a collaboration with Michelle Kendrick and Harrison Higgs, faculty in the Program in Electronic Media and Culture at Washington State University–Vancouver. By 1998, the original history of science fiction project had collapsed: The costs of producing a set of CD-ROMs dwarfed what the university could envision as a payoff in student fees for an on-line course.[1] The Mars "module" by then had assumed a life of its own. Backed by grants from West Virginia University and Washington State University, *Red Planet: Scientific and Cultural Encounters with Mars* (Markley et al. 2001) became the first scholarly-educational DVD-ROM authored from the ground up to be published by a major university press.[2] The authoring process took four years and forced the authors and designers to confront both the practical problems and the theoretical complexities presented by new media technologies that continually reshaped and were reshaped by our efforts.[3]

Between 1997 and 2001, the rapidly changing capabilities of hardware and software redefined the values and assumptions that informed the architecture of multimedia; in turn, the resulting changes in how we perceived the relationships among text, video, animations, sound, and static images profoundly challenged traditional and chic hypertextual accounts of content and form, text and visual design. In describing in roughly chronological fashion the problems we encountered and the work-arounds we adopted, we want to develop a larger theoretical argument: "Information architecture" and "theory" are never distinct concerns. Our experience in authoring *Red Planet* suggests that new media pose both problems—perhaps even the end of "alphabetic consciousness," as David Porush (1998) argues—and opportunities for scholars willing and able to invest the necessary resources of time, money, and labor to help (re)define what we might call the visualization of scholarship.[4] In this respect, *Red Planet* may serve as a case study in the ways in which "text" and "visual images" interact dialogically with the changing technologies—sound, video, and dynamic animation—that are always in the process of redefining the conceptual frameworks and practices of multimedia. Beginning in 1998, the authors did twenty or so presentations of our work in progress to academic, student, and software industry audiences, including demonstrations at the University of Oklahoma, the Mars Society, and the Smithsonian Institution. In addition, early versions of the title were used in courses at West Virginia University (science fiction) and Washington State University (Web design and multimedia authoring); responses, questions, and informal suggestions from these audiences were often incorporated in ongoing redesigns of the project. These experiences reinforced our belief that *Red Planet*'s dialogic structure has significant ramifications for understanding the complexities of educational and scholarly multimedia. Whatever the future of DVD-ROM technology, the processes of authoring *Red Planet* offered us a crash course in beginning to understand the ways in which multimedia technologies break down and reconfigure the boundaries among "scholarship," "pedagogy," and new media.

Our project began to take shape in a rapidly changing electronic environment. In the 1980s and early 1990s, information architecture was dominated by conceptions of hypertext in fiction, criticism, and software programs such as Jay Bolter's and Eastgate Systems' *StorySpace*, a program for creating hypertextual documents that could be organized in a multitude of possible ways.[5] Committed to the belief that hypertext could reproduce spatially the associative logic (they claimed) of the ways in which the mind worked, theorists such as Richard Lanham (1993) and George Landow (1992) posited that such an associative structure enacts a postmodern "freeing" of the reader to create her own narrative(s) by combining and recombining textual (and occasionally visual) elements. These theorists argued that hypertext's nonlinear architecture radically transformed temporal and spatial models of writing and reading. Because we read in time and therefore process information sequentially, hypertheorists argued, spatial nonlinearity implies a suspension or redefinition of a text's temporal and narrative structure. Espen Aarseth (1997), for example, maintained that "nonlinear" texts "have a positive distinction: the ability to vary, to produce different courses" (41–42). This definition, in effect, suggests that "linear" texts lack this desired "ability to vary"; instead, they exhibit a plodding and restrictive logic that can be read only *mono*logically. It was a short step, then, for some enthusiasts in the 1980s and early 1990s to argue that hypertext actualized the poststructuralist properties of language and narrative that they gleaned from readings of Roland Barthes, Jacques Derrida, and Michel Foucault, among others. For these humanists-turned-cybergurus, hypertext seemed poised to become both template and genetic code for the digital "revolution."

As early as 1994, however, Richard Grusin pointed out that such efforts were misguided because they misinterpreted the crucial insight of post-Saussurean theories of language: The written text was "deconstructive" long before digital media were conceived. In the same issue of *Configurations*, David Porush (1994) maintained that the archaeology of hypertext extended back to the Talmud and its dynamic models of language, interpretation, mind, and media. By recognizing that semantic meaning has never been unitary and therefore never can be reduced to mathematicized metaphors of "information," Grusin and Porush offered more theoretically sophisticated and historically accurate ways of understanding the complex relationships between text and visualization than the abstractions of hypertext. For all their claims to revolutionary sophistication, hypertext theorists modeled their conceptions of design on traditional Cartesian assumptions about space and time. "Nonlinearity," Aarseth (1997) argues, "as an alterity of textual linearity (monosequentiality), can be seen as a topological (rather than tropological) concept, in accordance with the principles of graph theory" (43). By making graph theory a constitutive metaphor for nonlinearity, he invokes an abstract, two-dimensional model of spatial relations by presupposing that time and "meaning" can encoded as semantic and logical "elements" within a geometric space. This implied logic renders discrete elements of both "content" (text) and "design" (image) as containers or counters that can be arranged and rearranged at will. Paradoxically, then, by insisting that hypertext allowed users to navigate among multiple narrative elements, theorists such as Lanham and Landow reinscribed a traditional model of communication theory—one at odds with the values and assumptions of a "postmodernism" that, they implied, could be identified with point-and-click interactivity. Although temporal and spatial elements can be arranged in any number of ways (well, in any number of ways within the algorithmic constraints of the code in which the software is written), what is *within* or *defined as* an "element" in a hypertext document effectively is blackboxed: Content is recast in metaphors of mathematical information. In fact, the decisions about what constitutes a textual or visual element actually *restrict* these content containers to a temporal or spatial unit (the sentence, the paragraph, the photograph) that reinforces traditional assumptions about language, meaning, and design.

Before beginning work on *Red Planet*, Markley (1993, 1–33) and Kendrick (1996) had voiced very different views of language, representation, and media theory. From the start, this title in the making was guided by the authors' distrust of the metaphor of text as "information" and the biases that made visualization a poor relation to textual content. In contrast to the values and assumptions of hypertext theory, multimedia produces dialogic interactions among and within textual and visual elements, if we understand "dialogic" in the complex ways theorized in Mikhail Bahktin's theory of language. Bahktin (1981) rejects the notion of a "unitary" or an authoritative language, a one-to-one correspondence between word and thing that seeks to repress the "process of historical becoming that is characteristic of all living language" (288). For Bahktin, each utterance is historically and socioculturally located; it penetrates and is penetrated by other utterances, "entangled, shot through with shared thoughts, points of view, alien value judgments and accents" (289). Marked by internally contested relations to other utterances, each utterance competes agonistically in "a dialogically agitated and tension-filled environment of alien words, value judgments, and accents" (276). Because *all* understanding is dialogic, every utterance—whether on a printed page or in a hypertext box—is internally contested. In this regard, hypertext does *not* actualize a "latent" dialogism within a previously "linear" or "rational" language but denies or represses the dialogic nature of *all* texts to further its own claims to be doing something revolutionary.

If words are always contextual and contested, then the introduction of static images, hot links, sound files, animation, and video raises the stakes exponentially by challenging the underlying metaphors of *information* that are invoked in most discussions of multimedia. As Aarseth's (1997) invocation of graph theory suggests, the concept of "information"—and the reduction of text and image to abstract, quasi-mathematical units—invokes a simplified view of how communication occurs. Because hot-link interactivity, as Kendrick (2001) argues, is limited by the algorithmic structure of computer code, the metaphors of mapping and associative logic that dominate discussions of hypertext-based multimedia are constrained by the physical and philosophical presuppositions of mathematical modeling that privilege "message" unproblematically over "noise." These metaphors depend on a conceptual model of information theory that presupposes both the transparency of the "code" and the unitary meaning and stability of the "message."

code
|
sender—> message—> receiver

For this model to work, the sender and receiver must share an unambiguous and fully understood code; the message must be articulated precisely within the semantic and logical constraints of the code; and the receiver must be able to reproduce the sender's intentions and meaning without any transcription errors, without "noise."

In his influential critique of communication theory, Michel Serres (1982) draws on the physical properties of thermodynamics to argue that no system—physical, electronic, or thermodynamic—transmits "messages" without "noise." If you hear static on a radio, boosting the signal increases the static in proportion to the volume of the music rather than eliminating it. This noise, Serres maintains, is not an "error" that can be rectified but an essential form of mediation that defines our comprehension of the signal. Within any communication system, the binary relation between sender and receiver, who must share a "code" in order to conceive and understand the "message," is always mediated by a parasite (*le parasite* is the technical term in French for electronic noise) that mediates, interrupts, and paradoxically helps to produce "meaning." It is only *against* such noise that "meaning" can take shape: "we know of no system," Serres notes, "that functions perfectly, . . . without losses, flights, wear and tear, errors, accidents, opacity—a system whose return is one for one, where the yield is maximal" (12–13). Such "losses," "errors," and "wear and tear" within communication systems are, in important ways, analogous to the complexity defined by Bahktin's dialogic description of language: media always generate a welter of effects, ongoing constructions and reconstructions, interpretations and misinterpretations, that the sender or speaker cannot control.

Crucially, as William Paulson (1988) argues, this "noise," these competing interpretations and meanings, is precisely what we term "literature," "history," "culture." In short, Serres and Bahktin offer a means to think against the grain of those philosophical assumptions that transform complex issues in the theory and practice of multimedia to the problems of surface, interface, and transparency that guided many responses to the widespread proliferation of digital media. Rather than "freeing" a disembodied user to "create" infinite meanings in an enabling digital universe, multimedia forces users to grapple with the historical and cultural embeddedness of technologies of representation. In this regard, multimedia cannot be defined romantically as a set of frictionless interfaces or as the metaphysical abstraction "cyberspace." Multimedia must be described heuristically and materially by complex networks: of workers, investors, programmers, producers, and consumers; of computer codes, cables, connections; of raw materials consumed to fuel ever-expanding demands for electricity; and of errors, updates, patches, and work-arounds. These interactions are both dynamic and irreducibly complex. Orchestrating static images, text, hyperlinks, audio, animation, and video dynamically does not allow us to transcend the noise of media culture, but it does allow us to turn up the volume.

The complexity of multimedia, according to Jay David Bolter and Richard Grusin (1999), can be understood as a consequence of "remediation." Following Marshall McLuhan, they argue that all media cannibalize and subsume previous technologies of representation. Whether we enhance media to achieve unprecedented standards of visual, auditory, or sensory quality (hypermedia) or design systems that try to remain as inconspicuous as possible (transparent media), the ultimate goal is the same: to reduce noise to a minimum and thereby "to get past the limits of representation and to achieve the real" (Bolter and Grusin 1996, 313). This desire to transcend representation, however, is shaped by perceptions of "reality" derived from media with which we already are familiar. As every designer recognizes, we can describe improvements in verisimilitude *only* by judging new media against the perceived inadequacy of those media that they are designed to replace. "Each new medium," Bolter and Grusin (1996) assert, "is justified because it fills a lack or repairs a fault in its predecessor, because it fulfills the unkept promise of an older medium [343]. . . . In each case that inadequacy is represented as a lack of immediacy" (314). Our sense of what is "new" or "revolutionary" in media—of what constitutes "immediacy"—is structured by perceptions of an always mediated reality, a reality, in other words, that is always in the process of becoming obsolete. Because multimedia is irrevocably dialogic, adding an image to a text, inserting navigation tools, enhancing rollover buttons, and, in short, enriching a mediated environment cannot eliminate noise or bring us closer to experiencing an "immediacy." Rather, multimedia transforms and recombines local, discrete elements, remediating previous technologies of representation. Rather than transcending technology, multimedia reinforces our sense of what Kendrick (1996) terms the "technological real": the recognition, whether implicit or explicit, that consciousness, identity, and "reality" are and always have been mediated by technology and that this mediation is always dynamic (144).

By the time we began *Red Planet* in 1997, hypertext had been superseded practically and in the public consciousness by the World Wide Web. The Web, like CD-ROMs, remediated and cannibalized hypertext documents as well as a variety of print media that redefined users' expectations about what information architecture "should" look like, appropriating layouts and designs from magazines, newspapers, video games, books, graphs, photographs, logos, billboards, and so on. Our expectations of verisimilitude are redefined continually by technological mediation: What looks "real" one year is second nature the next (that is, it exhibits an expected level of technological competence) and out-of-date the one following that. On the Web and in multimedia authoring more generally, technological improvements (upgrades for downloading capabilities, for example) continually up the ante in terms of users' expectations of what is cool, cutting edge, useful, and aesthetically pleasing. But no standard of verisimilitude—pixels per screen, processing speed, RAM, storage capacity, improvements in downloading time, design modifications—can ever render media transparent. If our desire for verisimilitude on the Web and in technologies of visualization drives expanding expenditures of time, money, intellectual and manual labor, and raw materials, it also forces developers to make decisions about what constitutes marketable standards of immediacy. In authoring *Red Planet*, we were caught in an ongoing process of having to decide what we could afford in time, money, and labor to live up to our grant application claims that DVD-ROM could do what neither CD-ROM nor the Web could manage: integrate hours of high-quality video into a scholarly multimedia project and, in the process, redefine heuristically the dialogic interactions among text, image, animation, and video. At the same time, we had to engineer downward the minimum requirements of RAM, operating systems, monitor resolutions, and so on to avoid pricing our product out of a "mainstream" educational market.

From the start, we conceived of *Red Planet* as the scholarly "equivalent" of a book, that is, as a thesis-driven work of original scholarship. We confronted the challenge of producing a multimedia cross-disciplinary title that would extend beyond models of the electronic textbook or reference work yet, paradoxically, would have to remediate the very forms of digital media that were shot through with the commercialized values, assumptions, and expectations of the Web. If, as Markley (1994) and Kendrick (2001) have argued (see also Lewontin 1991; Hayles 1991, 1993), the dominant models for hypertext educational materials were commercial e-undertakings, we were faced with the question of whether a scholarly DVD-ROM could appeal to audiences beyond specialists in planetary astronomy and critics of science fiction and still claim the cultural capital of a refereed publication. Although the title would include basic scientific information about Mars—key definitions, animations of its orbital mechanics, and the basics of planetary geology—it was not limited to a specific grade level as a "teaching tool" or, for that matter, to a single discipline. We were committed to harnessing multimedia to present a historical and theoretical argument: since the time of Schiaparelli and Lowell, Mars has been a site on which scientists, science-fiction writers, readers, and the "general public" have projected and repressed anxieties about ecological degradation on Earth. The scientific quest to find or rule out the possibility of life on the red planet, we argued, was part of a complex narrative of planetary evolution that motivated science-fiction writers since the nineteenth century to depict Mars as a "dying planet," the contested site of radically different views of humankind's relationship to a terrestrial environment on the verge of apocalyptic collapse.[6]

This decision to resist dumbing down the narrative to straightforward "information" of the sort that is readily available on NASA Web sites meant that, in addition to solving technical problems on a shoestring budget, the project had to distinguish itself from Web-based teaching materials so that *Red Planet* would be seen as the scholarly "equivalent" of a book, as a work of original scholarship. In contesting the default assumption that *all* electronic media had to be geared toward consumerist models of education, the four-year authoring of the DVD-ROM raised crucial questions about the financing and labor involved in a project of its size, complexity, and duration. In the absence of a mid-six-figure grant to fund the completion of *Red Planet*, the only way to compensate individuals who contributed to the authorial, artistic, and technical production of the title would come through the delayed forms of gratification that are valued and rewarded in academic institutions: publication or, for the designers, creative credit. Our insistence that *Red Planet* was *not* a CD-ROM was the result of a strategic and political decision about the ways in which all participants in the project—professors, graduate students, and undergraduates—perceived their roles within the contexts of academic labor and professional capital. If the project had been conceived of as an "instructional tool," an electronic reference guide that collected and arranged previously published and available information, then *Red Planet* could not claim much in the way of professional recognition. Without the promise of such credit, the project never would have been completed. Treating *Red Planet* as the equivalent of a book and, for the designers, a creative work influenced our decision to seek an academic publisher rather than a commercial purveyor of software or textbooks. In turn, the dialogic form of multimedia redefined the ways in which we came to think about "scholarship" and publication credit.

On a day-to-day, task-by-task basis, the production of *Red Planet* broke down conventional barriers between "authors" and "designers," between content and form. Most multimedia projects in the humanities and social sciences have reinscribed a division of labor between "content" and "medium." Although there are more than a few cases of humanists becoming enthralled by Web design and CD-ROM authoring, the majority of commercial educational software treats "content" as a given, reified as "information" that has to be encoded within a programming language and designed in such a way as to enhance its "usability."[7] In short, the division between textual labor (writing) and design labor (making it look good on the screen) reinforces divisions apparent elsewhere throughout the university and society as a whole between intellectual and manual work. With the exception of digital artists, those who write or patch code, who design your department's home page, and who service your e-mail account tend not to be "regular" faculty but either persons employed specifically to troubleshoot computer and network problems or an assortment of catch-as-catch-can graduate assistants, work-study students, part-time employees, or consultants hired piecemeal by the university. In such a universe, it should be no surprise to hear (as we have) colleagues say, "I have this great idea for a DVD-ROM; I'll write the text and give it to somebody who will do the design." The assumptions underlying this kind of comment not only reinscribe a division of labor but reinforce conventional divisions between form and content, media and message, hired help and "authors." Such biases are held by both humanists and designers: If a scholar produces "content," then the design is treated as packaging; if a publisher of electronic encyclopedias or textbooks produces a multimedia version of a remediated print text, then that "content" is black-boxed.[8] Writers write, artists design. Although professors may be encouraged to develop Web sites for courses, to set up list-servs for class discussion, most recognize that the time spent constructing such sites far outweighs any actual payback in terms of professional recognition, merit raises, and so on.[9] After much discussion, we found that we had to adopt the semiotics of both print and multimedia to accommodate the four primary and three secondary authors on a colophon page, as well as a more detailed and elaborate credits page to indicate the kinds of contributions made to content and information architecture by the authors and other individuals.

Our decision to produce *Red Planet* as a DVD-ROM had important consequences for how we conceived and continually modified its structure and design. *Red Planet* remediates self-consciously several generations (at least) of filmic and digital media: the minimalist, head-and-shoulders video inherited from television documentaries; hot-linked keywords derived from CD-ROM, and more distantly, biblical exegesis; digital artwork premised on studio collage techniques; and sophisticated animations. This ancestry carried with it a host of assumptions about how a software title should look, "feel," and behave. The dialogic design of *Red Planet*, in this regard, takes into account the unpredictability of user needs and demands by incorporating the elements of the commercialized Internet—flexibility, layered content, and intermediality (a dialogic interlacing of formats: video, audio clips, voice-over narration, 200 pages of text, hundreds of photographs, and diagrams)—while preserving a sense of narrative coherence.

By distancing our project from the metaphors of mapping and gaming that have guided hypertext and multimedia development, we tried to avoid sacrificing narrative coherence and intellectual rigor to consumerist desires for easy access to dumbed-down information.

One way that we sought to address the danger of dumbed-down information was to emphasize that static and dynamic images play crucial roles in providing a complex introduction to Mars as a scientific and cultural artifact, even as they entice media-savvy users by hypermediating video and animation. For instance, the hyperlinked animations of orbital mechanics (opposition, conjunction, retrograde motion) illustrate key concepts that a student or educated layperson would have to know to understand why windows for launches of spacecraft to Mars occur only once every twenty-six months. But these animations contribute to the historical and conceptual narrative, demonstrating why debates about life on Mars and the appearance of scientific articles about the planet cluster during the years of "good" oppositions, when Mars approaches as close as 34 million miles to Earth. The dialogic interplay between such dynamic images and the text, in this regard, becomes essential to the conceptual and artistic design of the overarching narrative. Scientific "information" is not slighted, and users still get to watch video clips of the devil girl from Mars zapping hapless Earth males.

Given its conceptual, historical, and generic argument, we assumed that our commitment to narrative would distinguish *Red Planet* from software conceived as chunks or modules. But we soon realized that narrative itself is entwined with the notions of space and visualization as well as that of time, and the ongoing challenge we faced was to find strategies of visualization that would supplement or enhance the narrative rather than disrupt it. One of the reasons *StorySpace* was so popular for a brief time in the 1990s was that it provided a visual *space* in which to define narrative relations: Textual elements existed in boxes that could be rearranged, hyperlinked, and illustrated. Aware of the vast amount of information that had to be organized, we structured the narrative (in contrast to the remediated print technology of hypertext) in a series of chapters that run chronologically from ancient observations to recent Mars missions, including two chapters on science fiction and its significance in shaping scientific and popular perceptions of the planet. Because users are able to access our Web sites directly from the DVD-ROM, and thus updates on missions, a bibliography, continuing debates, and links to other sites, *Red Planet* would have a flexibility that a scientific study of Mars or a literary-critical analysis of science fiction about the planet would lack.

In defining the shape of this cross-disciplinary narrative, we had to come up with visual ways to direct the user: The more data and options we had to consider, the more complex such design and conceptual decisions became. For early versions, the table of

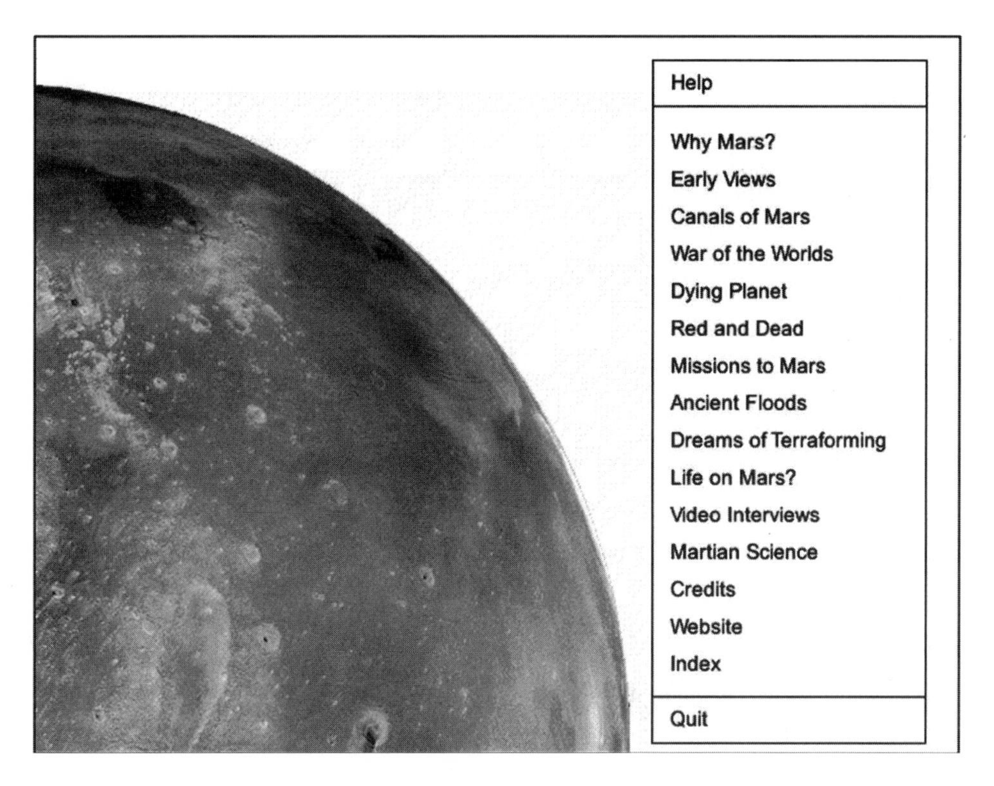

| **Figure 3.1** |

Table of Contents for *Red Planet.*

contents (figure 3.1) included chapter titles; but, prompted by the Marketing Department at the University of Pennsylvania Press, we added pop-up boxes for each chapter.[10] Each chapter heading features two rollover links; when a chapter is pointed to, the user sees a visual image that has an iconic relationship to the content of the chapter and a list of five to eight subheadings within the chapter: point to "Red and Dead," for example (figure 3.2), and the subheadings appear as a rollover. The rollovers remediate self-consciously the browser design of multimedia search engines, while emphasizing visually the location of information within the historical and conceptual narrative that the chapters describe.

These kinds of composing decisions indicate how design and content questions impinge on each other. The index, for example, borrows from both the book and the Web page; it identifies relevant screens and links directly to important concepts in the

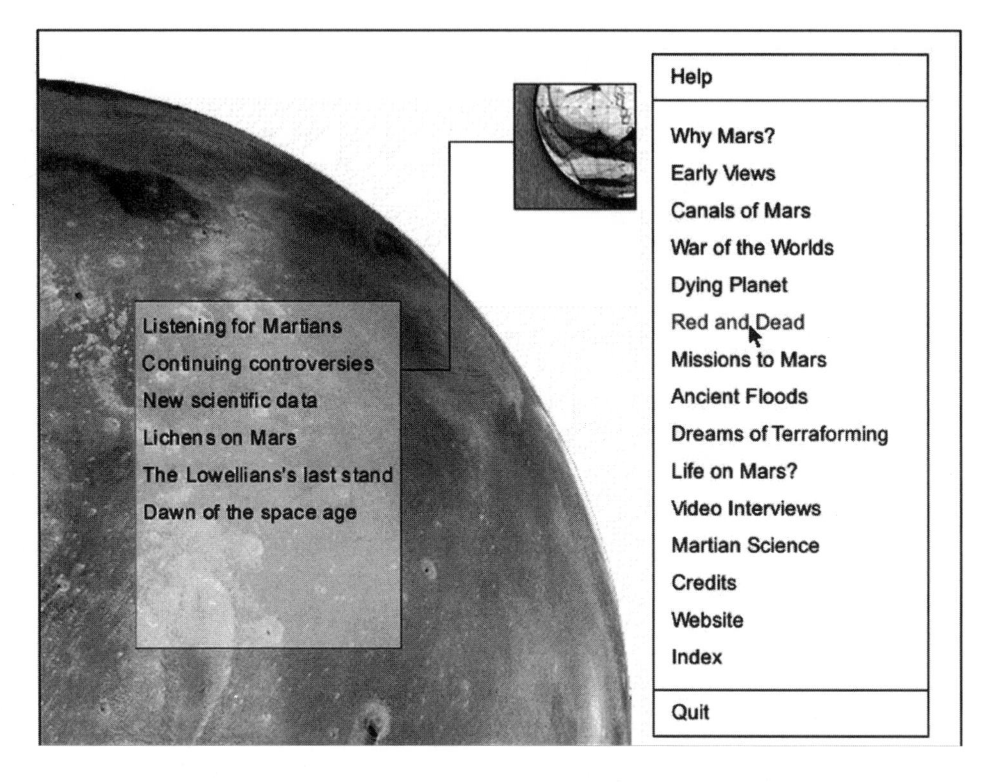

| **Figure 3.2** |

A rollover reveals detail about each section.

body of the work. In a variety of contexts, then, we had to come up with intuitive visual navigation strategies. *StorySpace* handled the problem of visual design with a floating box suspended above the text boxes and carved with cryptic signs (arrows and boxes). Working ten years after *StorySpace*'s peak in popularity and having the benefit of Web-savvy users, we were able to integrate the navigation bar into the project with an easily recognizable set of tools: forward and back buttons, a button for turning off sound, and a button for returning to the main menu. Eventually, the number of hyperlinked and pop-up screens (figure 3.3), used to define scientific terms, outnumbered the pages of the main narrative.

But these links, in different ways, were not ancillary to the main narrative, not analogous to footnotes, but explorations that continually redefined the shape of the narrative. The biography of Russian science-fiction novelist Alexander Bogdanov, for

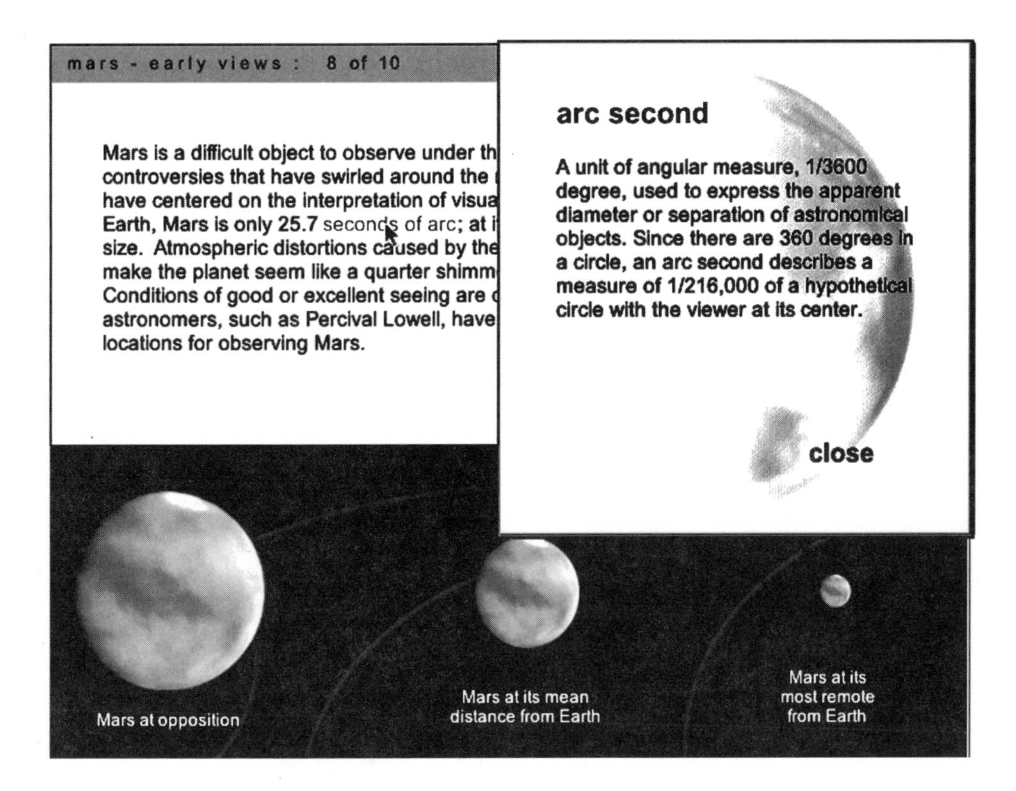

| **Figure 3.3** |

Screen featuring a "popup" glossary definition.

example, included the kind of information about the crucial role he played in the first Russian Revolution of 1905 that would be incorporated in a scholarly book (as it is in Markley's forthcoming *Dying Planet*) but that on the DVD-ROM can be skipped, skimmed, or used as a means to redefine the contours of the narrative. A science student trying to remember what "aphelion" means may skip Bogdanov and the two major science fiction chapters completely; a cultural critic may read through the Bogdanov screens carefully, note the Key References on early twentieth-century science fiction, check the extensive bibliography on our Web site, and, clicking twice, be able to locate or purchase key books that discuss Bogdanov, such as Robert Service's (2000) biography of Lenin.

Rather than describe the individual chapters in a blow-by-blow fashion, we want to concentrate on the significance of integrating video into *Red Planet* for three reasons:

Video presented us with many of our most daunting technical and design problems; video radically alters traditional art-historical conceptions of the image; and, given the development of DSL and broadband networks, high-quality video seems likely to define the future of multimedia. In large measure, our decision to include extended video clips in the DVD-ROM meant that conceptually as well as technically, we were breaking new ground for educational or scholarly multimedia. Even in 1997, we recognized CD-ROM was destined to be replaced by DVD-ROM. A single-sided DVD holds a minimum of 4.38 GB of data, a CD-ROM about one-seventh of that. This storage capacity means that a DVD-ROM can hold several hours of high-resolution video, the equivalent of a book ninety feet thick; as significantly, DVD-ROM reads data a rate seven times faster than CD-ROM, a significant difference when one is compressing video. Video on CD-ROM is a pixilated, impressionistic oddity; on DVD-ROM, video begins to achieve something of the semiotic verisimilitude we associate with film.

The video interviews on *Red Planet*, sixty of them, are lengthy clips (up to seven minutes) playing at fifteen frames per second, with a data rate of 300 KB per second. File sizes range from 4 MB (for short segments in the introduction, "Why Mars?") to 136 MB (Robert Zubrin's description of the Mars Direct Plan).[11] Files of this size effectively spell the end of CD-ROM as a storage medium in a video-rich, multimedia environment. Because we were working with proprietary software (Macromedia's Director, versions 5, then 6, then 7), we were able to use video boxes, 360 × 240 pixels, much larger than those on the Web, which measure 120 × 80 pixels. The video on *Red Planet*, excluding all other multimedia features (static images, animations, text, navigation bars, sound files, and so on), takes up 2.5 GB, the storage space of three CD-ROMs. Yet DVD-ROM allows for smooth, instantly accessible video that can be viewed and navigated easily. Video offered over the Internet, on the other hand, is usually pared down to ten and sometimes seven frames per second, with a data rate of 15 KB or lower per second. Video compressed for the World Wide Web, at this stage, consists of comparatively short, choppy clips or sound bites that do not allow for the kinds of in-depth, comprehensive responses offered by interviewees on *Red Planet*.[12] Even when video clips are squeezed into such small file sizes, many Internet users grow impatient with the time it takes to download clips. Using a 56K modem, it would take almost a week to download all the data on our DVD-ROM. Although cable lines obviously handle Internet video faster and more smoothly, they are expensive: The list price of *Red Planet* is $39.95, almost exactly the going rate in early 2002 for one month of high-speed Internet access.

From the start, we were confronted with the problem of how to integrate video interviews with NASA scientists, science-fiction authors, and cultural critics without interrupting the narrative. In this respect, the integration of video into a scholarly narrative called into question the subordination of image to hypertextual metaphors of mapping and problematized the notion of a linear narrative. Static images can be scanned and downloaded easily, as anyone who has constructed a Web site since 1995 realizes; video presented us with crucial decisions to sort through in negotiating among the remediated design and conceptual conventions of visualization: hypertext buttons, BBC talking heads, clips from 1950s science-fiction movies, navigation bars, hyperlinks, static images from a variety of print and electronic sources, and so forth.[13] In the days before video capture and editing became a standard feature on iMacs, we worked through a process of trial and error in deciding how video clips would enhance and complicate our historical and conceptual narrative. In early versions, the narrative (and voiceover) would stop; a video screen (360 × 240) would appear against a black or night sky backdrop and the talking head would talk. The video would end, and the next text and image screen would appear automatically. At this preliminary stage, the video clips functioned analogously to passages of primary material in a critical text: The previous screen introduced the "content" of the clip, the clip played, then the narrative continued. Other video clips in early versions required the user to point and click, to select an image of a talking head. The black backdrop and video screen would appear centered on the computer screen and disappear when the speaker finished. Automatically, the screen would return to the original page from which the user had linked. These video clips functioned, in this sense, by loose analogy to footnotes, "see also" entries in encyclopedias, and Web-based hyperlinks.

Our use of video, then, had a very different function from newscast or documentary sound bites. Television documentaries, for example, are limited by time and financial constraints; the clips must fit into a narrative at once entertaining and generically informational. Video must be integrated as seamlessly as possible into a medium that provides the viewer with limited options: watch, change channels, hit the mute button, or mow the lawn. In contrast, the video links that we developed were situated within a variety of spatial and temporal contexts. Because users can click on a video clip, turn it off by closing the window, replay it, or skip forward or backward in the narrative, the use of video becomes a dialogically fraught element: It enhances, disrupts, complexifies the notion of narrative itself. Significantly, it provides a visual semiotics that cannot be reduced to mere "information."

Our sense of narrative, in brief, had to adapt to the experience of having to integrate a variety of views, often radically at odds with each other, into our argument. To complicate matters, the interviews we used were conducted over a three-year period (1997–2000) at various locations (Denver; Los Angeles; Flagstaff, Arizona;

Morgantown, West Virginia; Toledo, Ohio) using different videographers and different high-end tape formats (3/4-inch, Beta, mini-dv).[14] Many of the tapes eventually had to be dubbed into mini-dv, although in 1997 and 1998 we were still renting expensive decks to do our own dubbing to and capturing from 3/4-inch tape. The majority of interviews were conducted before the failure of the 1999 missions (when the Mars Climate Orbiter went off course and burned in the Martian atmosphere and the Polar lander crashed: both failures were the result of human error); before all of the 60,000 photographs were taken by the Mars Global Surveyor, some with resolutions as precise as one meter per pixel; before the paradigm-bending announcements that sedimentary layers and recent evidence of liquid erosion on the Martian surface had been photographed; and before the February 2001 article that apparently confirmed the existence of nanofossils in Martian meteorite ALH84001 (Friedman et al. 2001). The questions we posed—and could pose—in interviews changed over time, and the interviewees' responses had to be selected, edited, tweaked to filter out sound distortions, and located within the narrative.

Our theoretical concerns with dialogic nature of multimedia became particularly relevant in the arrangement of the video interview section of *Red Planet*. The gateway screens of this section (figure 3.4) include the question posed to each scientist, critic, or author and feature head shots of each interviewee. Users are able to select the video interviews they want to watch and the order in which they want to watch them. The control bars allow users to stop, speed through, and go back to any point in the interview. Unlike the chapters, there is no narrative *other* than the unscripted and often animated responses of those interviewed: Jeff Moore, Carol Stoker, Chris McKay (all from NASA-Ames), Kim Stanley Robinson (the science-fiction novelist), Molly Rothenberg (Tulane), Richard Zare (Stanford), Robert Zubrin (President of the Mars Society), and Philip James (Toledo).[15] Video becomes the dominant content at this point, providing a dialogic forum through which users may investigate competing narratives about the future of Mars exploration, the benefits and dangers of terraforming another planet, and the significance of finding definitive evidence of past or future life on the red planet. It is left up to the user, with some help from the Educational Resources section, to analyze and, in effect, to *narrativize* the material. In every case, the physical appearances, voice tones, eye movements, facial expressions, and gestures of the interviewees become crucial factors in the user's process of making sense of their responses. Zubrin's enthusiasm, Rothenberg's skepticism, and Stoker's forceful justification for the continued exploration of space cannot be reduced to textual statements. The semiotic complexities of the users' responses to individuals discussing complicated scientific and sociocultural issues inflect the interviews with an immediacy, and

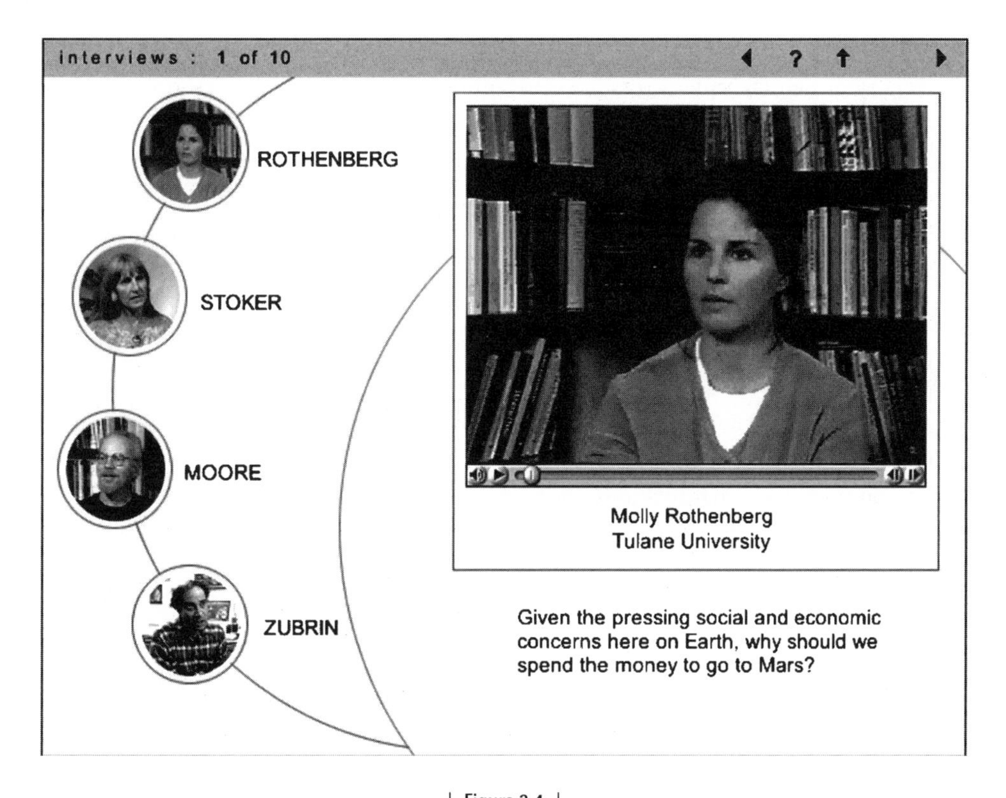

| Figure 3.4 |

Video interview screen: Each question is "asked" to a number of experts, and the clickable video answers grouped together on one screen.

a level of interpersonal indeterminacy, that would be lost if these experts' views were printed as a transcript or condensed to sound bites.

One of the effects of working with a DVD-ROM that uses video clips of three, four, and five minutes, then, is that the informational metaphors that dominate educational multimedia break down. One the one hand, we have known since the 1960 presidential debate that "information" includes a range of semiotic responses beyond what a text "says": Richard Nixon's five o'clock shadow, rather than his stand on civil rights, in retrospect, helped to elect John F. Kennedy. Yet despite decades of experience with Teflon presidents, media messaging, and rehearsed "debates," the default assumptions in multimedia still privilege text over (video) image. One of McLuhan's key insights—that most people respond to innovative technologies with the values, assumptions, and expectations derived from old technologies—can be illustrated by an experience we had with *Red Planet*. At our final demonstration of an early version of the title for the University of Pennsylvania Press prior to signing a publishing contract, one member of the publications board (a professor of communications, no less) interrupted us after about ninety seconds—midway through the three short video clips that introduce the title—to declare that talking heads were *not* multimedia. After we showed him other features—hot links to Key References, definitions of scientific terms, and extended segments of video and animation—he still maintained that what we were producing was an electronic coffee table book. In effect, he discounted any informational content in the video itself, at one point suggesting that a single adverb—"he said enthusiastically"—could do the same "work" (his term) as the video clip of Richard Zare narrating the chance sequence of events and personal favors that led to his using mass laser spectroscopy to discover the nanofossils in Martian meteorite ALH84001. Finally, we clicked on the screen that includes two audio clips from the Mercury Theatre's original 1938 broadcast of Howard Koch's adaptation of H. G. Wells's *War of the Worlds*. On this particular screen, no critical commentary or apparatus appears, although we later added three pages of links to display quotations from listeners who heard the broadcast and thought that Martians had indeed invaded New Jersey. "Oh," he said, at last convinced, "now that's multimedia." His default assumption seemed to be that multimedia was defined not by content but by the layered remediation of prior technologies.

In less than a decade, the World Wide Web has gone through several evolutions of transmission capabilities: from Ethernet, base 10 and 100, to DSL, to the fiber-optic networks of Internet 2. If this increase in network data rates heralds a coming remediation of previous "passive" media—television, video, film—as "new" interactive or immersive technologies, it also creates significant problems for humanists. The ideology of commercial viewing has had profound influences—often negative ones—on the ways in which students, teachers, software designers, and others perceive the relation between word and static image. It is likely that the next incarnations of the Web will make video close to ubiquitous on commercial sites. The economics of video-rich multimedia production, without question, will ratchet up development and authoring costs: Standard permissions for commercial film, for example, run more than $1,000 per minute. The vast majority of available grant money is earmarked for major supercomputing projects. The National Science Foundation initiative in Information Technology Research (NSF 00-126) in January 2001 defines three classes of projects: "Small projects" with total budgets up to $500,000; "Group projects," up to $5,000,000, with annual budgets of up to $1,000,000; and "Large projects" with total budgets up to $15,000,000 and annual budgets up to $3,000,000 (National Science Foundation 2000). In such an environment, writing in and for multimedia will have to assert its "traditional" values in an increasingly competitive and commercialized marketplace of high-quality images. The challenge of DVD-ROM and Internet 2 technologies, it seems, will be to reassert the value of "content development" so that educational titles do not become ghettoized.

We want to emphasize, however, that DVD-ROM is very much a temporary storage device, one, no doubt, that will be superseded. Given the current generations of computers, it is a reliable and inexpensive medium rather than a be-all and end-all for educational multimedia. DVD-ROM does not offer a replacement for Hollywood-quality entertainment, but it stands a good chance of defining a benchmark for video quality in digital education. *Red Planet* is not, as we constantly must tell people, a disk that can play in a standard DVD player. Neither is it a model to be emulated so much as it is a historical document, a means to think through the scholarly and professional legitimation of video and visual information. "So our virtues," says Aufidius in Shakespeare's *Coriolanus*, "lie in th'interpretation of the time." Visual literacy and digital technologies might not spell the end of alphabetic consciousness and scholarly argument, but they promise to redefine it, we hope, in ways that contest consumerist complacency and cyber-illiteracy. The dialogical complexity of multimedia necessarily complicates conceptions of education based on the transmission of information and, in the process, suggests other models of assessing the significance of new media besides consumer surveys and standardized test results. The irony of new media may be that the more universities and professors commit to next-generation networks, the more valuable Latour, Bahktin, and Serres may become in contesting the naive enthusiasms that have animated too many accounts of the digital "revolution."

The test of our dialogic description of information architecture in new media, however, may well come in the authoring of our next project, an educational DVD-ROM aimed at teaching fundamental concepts in physics to students while also emphasizing the social and cultural context of these formulas and methods. *Red Planet* has the benefit of being able to subsume cultural "content" and literary elements into a chronological narrative. In contrast, *The Gravity Project* (its working title) (Markley et al. forthcoming) is not anchored by such a narrative; in its current state, it incorporates a historical narrative—the development of theories of gravitation from Aristotle, to Descartes, to Kepler, to Newton, to Einstein—into extended explanations of basic principles that, in turn, depend heavily on videotaped experiments and interactive exercises. Remediating the textbook may not be as sexy as a leather-clad Martian dominatrix, but our attempt to secure funding to complete this DVD-ROM may well indicate whether dialogism sells in the conservative field of science education.

Initially we experimented with a "graphlike" architecture, with a timeline along the top of the screen and scientific principles arranged vertically on the left-hand side. But this design raised questions about the relationships among a coherent narrative (the historical development of gravitational theory), the need to inculcate basic scientific principles in students' minds, the desire to contextualize experiments and principles, and the pedagogical advantages of including extensive video clips in which historians of science describe the scientific and cultural significance of debates about the nature of gravity. As a heuristic model, we are not graphing hybrid structures into what we hope will be a coherent whole: instructional modules, each with set of first principles, experiments and exercises, and a larger historico-cultural narrative. The modules are ordered chronologically, each module beginning with a "gateway" page that gives the user four ways to approach the material.

The Gravity Project, with its four-way structure, promises to extend and remediate the experiences we have had with *Red Planet*. In particular, our organizing structure of exercises, experiments, first principles, and cultural history may dialogically disarrange themselves with each student's potential interaction. Quite clearly, the organization we have chosen presents its own dialogic interference patterns that are always and already forcing us to question both the form and content of the DVD-ROM: How do we decide what is an experiment and what an exercise? What is a first principle and what is an important cultural context? *Red Planet*, in this respect, is only the first step in a movement toward a more dialogically organized—and hence deorganized—theory of information architecture. It preserves features that are derived from scholarly books, such as the index (figure 3.5), but redefines them so that they take advantage of web-like interactivity. Clicking on an item in the index takes the user to the relevant screen or screens, and the toolbar then offers the option of continuing within the main narrative in that section or returning to the index. At this juncture, modifying the search functions to mimic those of the Web is a possibility for titles currently in development: The ultimate decision will depend on resolving technical, conceptual, and design issues and rethinking the decisions we made in authoring *Red Planet*. For now, *The Gravity Project* is an experiment in the ongoing evolution of multimedia culture. The improvisational natures of narrative, information architecture, and the dialogics of new media guarantee that this project, like *Red Planet*, never truly will be finished. Multimedia, like its print technology ancestors, must *be* rather than *mean*.

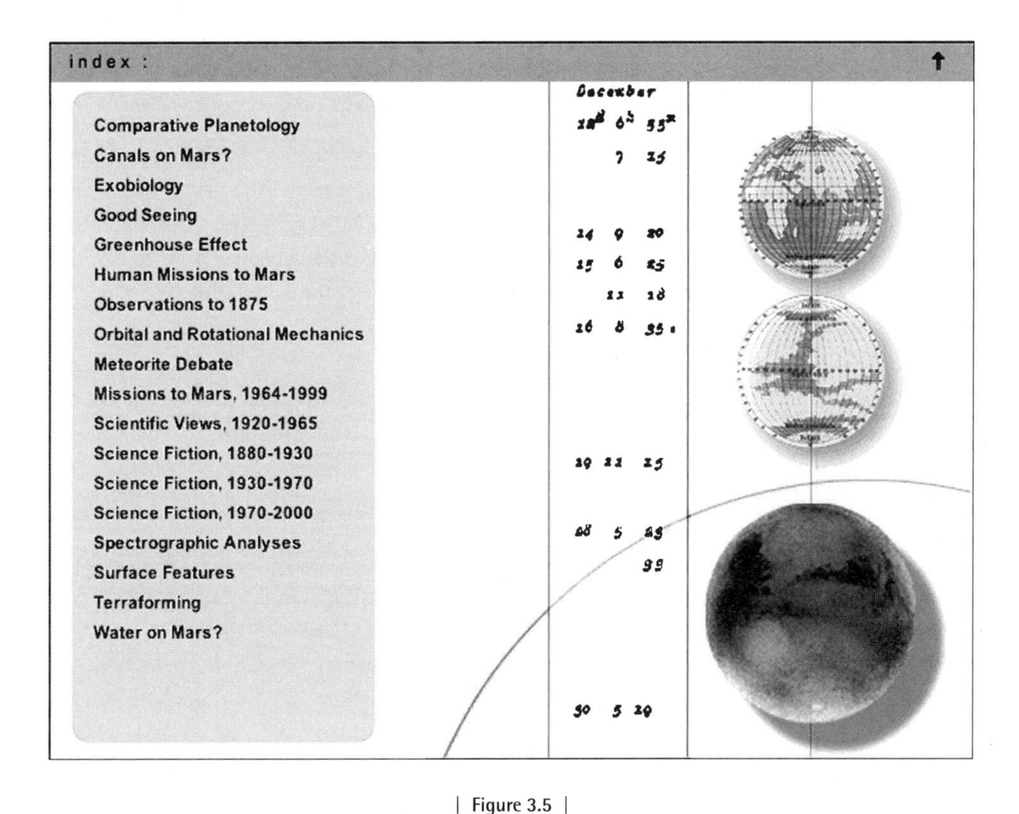

| Figure 3.5 |

Index main screen: Each category reveals subcategories that can be clicked to move to the appropriate screen.

Acknowledgments

We would like to acknowledge the Lowell Observatory for permission to use images from its archives and to the Wade Williams Collection, Corinth Films, for permission to use video clips from *Rocketship X-M* and *Devil Girl from Mars*.

Notes

1. The total development costs for multimedia authoring on CD-ROM in 1996 averaged $400,000: $100,000 for content creation and acquisition, $200,000 for production, $60,000 for testing, and $40,000 for overhead and miscellaneous costs (see Rosebush 1996, 259–275). Extrapolating these costs for a hypothetical DVD five

years later is difficult. Although the industry standard DVD-ROM holds 4.38 GB, seven times that of a CD-ROM, authoring software has become more sophisticated, and production costs, particularly for video processing, have become less expensive. A best-guess for a DVD-ROM produced *commercially* (that is, paying going labor rates of $100 to $300 per hour) would approach seven figures. The best source of information on DVD technology can be found at a site called Videodiscovery (2001).

2. An on-line preview of this multimedia DVD-ROM and the Mariner10: Educational Multimedia series of which it is part is available at <http://www.mariner10.com>.
3. Because the seven coauthors and others contributed their labor for academic capital (merit pay, tenure and promotion, summer stipends, and travel money), estimating a total cost for producing *Red Planet* depends on how one calculates percentages of research time spent on the project. A good guess is $250,000 (about half in grants and equipment from West Virginia and Washington State Universities). For background on theoretical issues in multimedia authoring, see Coyne 1995 and Chang, Eleftheriadis, and McClintock 1998.
4. Multimedia titles, as we discovered, are far more labor intensive than books. Markley estimates that for every hour he spent writing his book about Mars, he spent three to five hours working on his part of the collaborative project. In addition to research writing, his tasks included everything from learning the basics of multimedia editing and videography to grant writing.
5. There is an extensive bibliography on hypertext, cyberspace, and electronic environments. See, for representative examples, Brook and Boal 1995; Heim 1994; Laurel, 1993; Murray 1997; Ryan 2001; Stone 1996; and Woolley 1993. For examples of electronic hypertexts and virtual environments, see Laurel, Strickland, and Tow 1994; Greco 1995; Jackson 1999; and Strain and Vanhoosier-Carey 2001.
6. This argument is extended in Markley forthcoming.
7. As Kendrick (2001) argues, this conception of interactivity is often reduced to the operations of pointing and clicking and measures of design efficacy to consumerist models of accessing predigested nuggets of information. See also Nielsen 1999.
8. Tellingly, the largest grants to centers in the humanities have been geared toward digitizing existing materials: Johns Hopkins University Press Project Muse, on-line journals by subscription, the e-book Center at the University of Virginia. Thousands of literary texts are available from the center that can be downloaded as books using Microsoft Reader; the translation of print to a digital medium that mimics the physical appearance of the book seems to have matured to the point where there is not a lot left to do with or to texts. On the significance of visualization, see Stafford 1996.
9. Although materials archived on Web sites have become the center of contention about the "ownership" of electronic media, they frequently serve as glorified handouts. Although two of the principals on *Red Planet*, Higgs and Kendrick, teach Web design, Web sites themselves seem insufficient evidence of scholarly activity.

10. For the first two years, all production work took place in the Multimedia Application Research Studio lab at Washington State University–Vancouver. In 2000 and 2001, some additional authoring work (particularly video and Web design work) were undertaken at West Virginia University. Markley, Burgess, Daniel Tripp, Hamming, and Catherine Gouge made some two dozen trips to Vancouver or interview locations during the production process. Each trip resulted in another "version" of the title. All told, then, there were probably at least twenty provisional versions that tried, discarded, and refined a variety of design options.
11. Zubrin is president of the Mars Society and a frequent talking head on educational television. His plan for a low-cost, manned mission to the red planet is outlined in Zubrin 1996.
12. Christopher Breen (2000) recommends the following standards for digital video: CD-ROM, resolutions in pixels: 320 × 240; frames per second: 15; data rate: 100 KB per second. For the Web: 192 × 144; frames per second: 7.5; KB per second: 5.
13. Because many of the images used in *Red Planet* were from NASA and therefore in the public domain, we could draw from a substantial archive.
14. Each video shoot cost us between $250 and $500. In addition, travel expenses to various sites (or to have some interviewees come to us) ate up large amounts of our grant money.
15. Additional video interviews with Katherine Hayles (UCLA) and Henry Giclas (Lowell Observatory) are included in the body of the narrative.

Works Cited

Aarseth, Espen. 1997. *Cybertext: Perspectives on Ergodic Literature*. Baltimore: Johns Hopkins University Press, 1997.

Bahktin, Mikhail. 1981. *The Dialogic Imagination*, trans. Michael Holquist and Caryl Emerson. Austin: University of Texas Press.

Bolter, Jay David, and Richard Grusin. 1996. "Remediation." *Configurations* 5:308–350.

Bolter, Jay David, and Richard Grusin. 1999. *Remediation: Understanding New Media*. Cambridge: MIT Press.

Breen, Christopher. 2000. "How to Wrap Your iMovie." *Mac World* (December):66–70.

Brook, James, and Iain A. Boal, eds. 1995. *Resisting the Virtual Life: The Culture and Politics of Information*. San Francisco: City Lights.

Chang, Shih-Fu, Alexandros Eleftheriadis, and Robert McClintock. 1998. "Next Generation Content Representation, Creation, and Searching for New-Media Applications in Education." *Proceedings of the IEEE* 86:884–904.

Coyne, Richard. 1995. *Designing Information Technology in the Postmodern Age: From Method to Metaphor*. Cambridge, MA: MIT Press.

Cunningham, Steve, and Judson Rosebush. 1996. *Electronic Publishing on CD-ROM*. Sebastapol, CA: O'Reilly.

Friedman, E. Imre, Jacek Wierzchos, Carmen Ascasao, and Michael Winklhofer. 2001. "Chains of Magnetite Crystals in the Meteorite ALH84001: Evidence of Biological Origin." *Proceedings of the National Academy of Sciences* 98:2176–2181.

Greco, Diane. 1995. *Cyborg: Engineering the Body Electric*. Hypertext. Watertown, MA: Eastgate Systems.

Grusin, Richard. 1994. "What Is an Electronic Author? Theory and the Technological Fallacy." *Configurations* 2:469–483.

Hayles, Katherine. 1991. "Constrained Constructivism: Locating Scientific Inquiry in the Theater of Representation." *New Orleans Review* 18:76–85.

Hayles, Katherine. 1993. "Virtual Bodies and Flickering Signifiers." *October* 66 (Fall):69–91.

Heim, Michael. 1994. *The Metaphysics of Virtual Reality*. New York: Oxford University Press.

Jackson, Shelley. 1999. *Patchwork Girl*. Hypertext. Watertown, MA: Eastgate Systems.

Kendrick, Michelle. 1996. "Cyberspace and the Technological Real." *Virtual Reality and Its Discontents*, ed. Robert Markley, 143–160. Baltimore: Johns Hopkins University Press.

Kendrick, Michelle. 2001. "Interactive Technology and the Remediation of the Subject of Writing." *Configurations* 9:231–251.

Landow, George. 1992. *Hypertext: The Convergence of Contemporary Critical Theory and Technology*. Baltimore: Johns Hopkins University Press.

Lanham, Richard. 1993. *The Electric Word: Democracy, Technology, and the Arts*. Chicago: University of Chicago Press.

Laurel, Brenda. 1993. *Computers as Theater*. Boston: Addison-Wesley.

Laurel, Brenda, Rachel Strickland, and Rob Tow. 1994. "Placeholder: Landscape and Narrative in Virtual Environments." *Computer Graphics*, 28:118–126.

Lewontin, Richard. 1991. "Facts and the Factitious in the Natural Sciences." *Critical Inquiry* 18:140–153.

Markley, Robert. 1993. *Fallen Languages: Crises of Representation in Newtonian England, 1660–1740*. Ithaca: Cornell University Press.

Markley, Robert. 1994. "Boundaries: Mathematics, Alienation, and the Metaphysics of Cyberspace." *Configurations* 2:485–507.

Markley, Robert. Forthcoming. *Dying Planet: Mars and the Anxieties of Ecology from the Canals to Terraformation*. Durham: Duke University Press.

Markley, Robert, Harrison Higgs, Michelle Kendrick, and Helen Burgess, with Jeanne Hamming, Jeanette Okinczyc, and Daniel Tripp. 2001. *Red Planet: Scientific and Cultural Encounters with Mars*. DVD-ROM. Philadelphia: University of Pennsylvania Press.

Markley, Robert, Earl Scime, Helen Burgess, Jeanne Hamming, and Daniel Tripp. Forthcoming. *The Gravity Project.* DVD-ROM. Philadelphia: University of Pennsylvania Press.

Murray, Janet. 1997. *Hamlet on the Holodeck: The Future of Narrative in Cyberspace.* Cambridge: MIT Press.

National Science Foundation. 2000. Initiative in Information Technology Research (NSF-00-126), January 2001. Available at <http://www.nsf.gov/pubs/2001> (accessed December 10, 2000).

Nielsen, Jakob. 1999. *Designing Web Usability: The Practice of Simplicity.* Indianapolis: New Riders.

Paulson, William. 1988. *The Noise of Culture: Literary Texts in a World of Information.* Ithaca: Cornell University Press.

Porush, David. 1994. "Hacking the Brainstem: Postmodern Metaphysics and Stephenson's *Snow Crash.*" *Configurations* 2:537–571.

Porush, David. 1998. "Telepathy: Alphabetic Consciousness and the Age of Cyborg Illiteracy." In *Virtual Futures: Cyberotics, Technology and Post-Human Pragmatism,* ed. Joan Broadhurst Dixon and Eric Cassidy, 45–64. London: Routledge.

Ryan, Marie-Laure. 2001. *Narrative as Virtual Reality: Immersion and Interactivity in Literature and Electronic Media.* Baltimore: Johns Hopkins University Press.

Serres, Michel. 1982. *The Parasite,* trans. Lawrence Schehr. Baltimore: Johns Hopkins University Press.

Service, Robert. 2000. *Lenin: A Biography.* Cambridge: Harvard University Press.

Stafford, Barbara Maria. 1996. *Good Looking: Essays on the Virtue of Images.* Cambridge: MIT Press.

Stone, Alluquere Rosanne. 1996. *The War of Desire and Technology at the Close of the Mechanical Age.* Cambridge: MIT Press.

Strain, Ellen, and Greg Vanhoosier-Carey. 2001. *Griffith in Context: A Multimedia Exploration of D. W. Griffith's* The Birth of a Nation. Available: <http://www.griffith-in-context.gatech.edu/> (accessed June 12, 2001).

Videodiscovery. 2001. <http://www.videodiscovery.com/vdyweb/dvd/dvdfaq.html>, accessed on May 30, 1999.

Woolley, Benjamin. 1993. *Virtual Worlds: A Journey in Hype and Hyperreality.* London and New York: Penguin Books.

Zubrin, Robert, with Richard Wagner. 1996. *The Case for Mars: The Plan to Settle the Red Planet and Why We Must.* New York: Free Press.

II

Historical Relationships between Word and Image

| 4 |

RECOVERING THE MULTIMEDIA HISTORY OF WRITING IN THE PUBLIC TEXTS OF ANCIENT EGYPT

Carol S. Lipson

Introduction

Increasingly the academic world has been recognizing the turn to the visual in our culture (Bolter and Grusin 1999; Kress 1998; Mitchell 1994; Seward 1997; Stafford 1996). Within composition and rhetoric programs, the trend is welcomed by some, viewed as opportunity by some, and ignored or deplored by others. Specialists in text often harbor a mistrust of images, a fear that the advent of graphics will negatively affect emphasis on sustained rational argument or on the sophisticated critical reflective thought that they value in student development. Few faculty members in composition or rhetoric have received training in multimedia composing or in dealing with graphics in general, yet many realize the need to acquire such expertise within a writing program. The National Council of Teachers of English (NCTE) in 1996 passed a resolution on the need for visual literacy, announcing in the background section that "teachers should guide students in constructing meaning through creating and viewing nonprint texts." The NCTE resolution fully recognizes that such an initiative requires professional development on the part of teachers to prepare them for handling visual literacy in the classroom.

At the same time, the impact of postmodernist theories in the academy has helped foster a recognition of the dominance of the Western model of communication, with its origin in Greece, and the effect on other cultural groups whose communication practices and values differ considerably. Many feminists, for instance, call for modes of academic and scholarly writing that do not privilege the argument and its aggressive logic of attack. The present situation seems to offer an opportunity to create forms of multimodal texts that allow for the making of meaning in ways that prove amenable to a wide range of cultural groups. To understand the effects of any new multimedia forms that might be developed, however, we need to understand better the ways that language interacts with other media. Such interaction is by no means a new phenomenon but is in fact fundamental to one of the earliest systems of writing: that of ancient Egypt. In the fields of composition and rhetoric, the inherent multimodal aspect of the history of writing has effectively been silenced. As we begin to address the relation of text and visuals in our own culture, it behooves us to begin examining the historical development of such multimodal representations.

Recent work by Jay Bolter and Richard Grusin (1999) examines historical developments in print, visual, and electronic media, beginning with the Renaissance. Bolter and Grusin look at the ways new media repurpose older media, and they usefully complicate oversimplified categorizations of the differences between visual communication and textual communication (see Seward 1997; Schlain 1998) by directing attention to the existence of very different styles and strategies of approach to visual media. My project here extends this historical gaze considerably to very early stages in the history of writing in ancient Egypt. I direct attention to some fundamental ways that the early literate culture of Egypt developed distinctive principles and approaches for communicating with visuals and with text and to ways that the principles crossed media, applying similarly to both graphics and verbal elements. Italian art theorist Mario Perniola (1995) has made a suggestive claim that ancient Egypt created a vast combinatorial system of interchangeable media elements. According to Perniola's claim, principles were created for combining elements of interchangeable media: text, graphics, architecture, and ritual. Perniola does not, however, illustrate or demonstrate this claim. Here I wish to look closely at aspects of the Egyptian treatment of text and visuals in relation to one another.

For this analysis, I will examine public monuments from ancient Egypt covering a span of approximately fifteen hundred years, from the very beginnings of writing in the predynastic period (3100 BCE) to the eighteenth dynasty of Egyptian pharaohs (1450 BCE). Although a simpler form of writing was developed in Egypt for letters, records, and literary papyri, the public monuments, all carved in stone, continued to utilize the more elaborate and artistic hieroglyphs. All writings and art in ancient Egypt came from the elite; in fact, all were commissioned and created by members of the state bureaucracy. The public monuments functioned within the umbrella of official state rhetoric, serving important rhetorical purposes within the culture. They are strongly epideictic. These public texts served as performances, as utterances conveying ideological and cultural values.

My work here recovers the presence of the visual in the early history of writing. To do this, I apply a system for analyzing the visual and its interplay with text. Specifically, I will build on Robert Horn's (1998) *Visual Language*, in which he introduces a way of looking at visual concepts in relation to language concepts. He suggests the existence of a new visual language, incorporating both images and texts, which is analyzable in terms of syntax, semantics, and pragmatics. Though he often emphasizes the newness of this language, he also acknowledges the close connections between this new language and the ancient Egyptian hieroglyphic system (25). Yet his entire discussion of the visual language of ancient Egypt encompasses a mere two pages. I will use his analytical approach to flesh out a more detailed sense of the ways the visual language of ancient Egypt conveyed its meanings.

In his analysis, Horn (1998) defines syntax as "the study of how the basic elements can be identified and combined into units" (54). In language, the basic foundational elements are morphemes and words. In visual language, elements are also combined into wholes, and Horn focuses our attention on the ways that smaller units are integrated into larger ones. Spatial relations become telling points of syntactic and semantic meaning (73). Horn turns to the six general approaches to visual groupings that Gestalt psychologists in the early twentieth century identified (75), using these to form the foundation of a visual language syntax. These general principles identify proximity, similarity, common region, connectedness, directional continuity, and closure as principles that afford meaning to visual structures and to combinations of graphics and language. For instance, we perceive elements placed in proximity to one another as having some relationship to one another. We perceive relationship in elements placed in a common region or in a closed area. We perceive a relationship in elements presented in a similar fashion. Thus particular fusions of syntactic elements become semantically meaningful. Relative size becomes an important signal of meaning, as does direction of gaze and what gets placed together in boxes or chunking units. In Horn's treatment, image-text combinations can have rhetorical functions, which involve persuasive impact, organization, and navigation. I will apply Horn's analysis to begin examining the distinctive public objects combining images and text that the ancient Egyptians created.

Multimodal Objects in Ancient Egypt: Relations between Writing and Art in the Narmer Palette

To illustrate the beginning stages of Egyptian image-texts, I will look at the Narmer Palette, one of the first available cultural artifacts in Egypt that employs hieroglyphs. This is a slate tablet, just over two feet long, in the form of a cosmetic palette used for applying makeup. In Egypt, both sexes used makeup around their eyes, and royalty would have such makeup applied by others. Cosmetic palettes are common in graves of the predynastic period of Egyptian history; many have elaborate carvings on both sides, as does the Narmer tablet.

Speculation abounds as to how such an object would have been used, what audience it would have, and for what purpose it was intended. Given that it is over two feet long and made of stone, and thus not very portable or moveable, it is hard to imagine that it was meant for actual use as a cosmetic palette; thus the palette is generally considered ceremonial or monumental. There is some suggestion that the palette was a votive object, used in religious ritual in a temple (Hornung 1992; Spencer 1993). But just how it was used remains a mystery. It was found in a large grave collection along with other prestige objects, spanning much of the late predynastic period, from the reigns of a range of kings. Besides palettes, the collection also included decorated maces, and all seem to focus on similar themes: the king as hunter, taming wild animals, or the king as victor in battle, with a conquered enemy figure. The latter focus is what we see on the Narmer Palette.

The Narmer Palette and Spatial Syntax

As we can see in figure 4.1, the surface on each side of the Narmer Palette is divided by horizontal carved lines into separate parts with subparts. The horizontal lines that divide the tablet are commonly called register lines. The division is hierarchical. On the back side of the palette, which is the left side of figure 4.1, the three parts created by the register lines resemble an inverted triangle, with the divine realm on top, the king and the human realm below that, and the defeated enemy at the bottom.

The top portion on both sides of the palette presents the divine realm, with the cow goddess, Hathor, framing a rather grand palace-like structure. The framing of the palace within representations of the divine realm associates the king with the divine. The spatial treatment suggests an equivalence: The king too is a god. In addition, two symbols are enclosed in the palace: One depicts a fish, specifically a catfish, known as an aggressive survivor in later texts; the other symbol depicts a chisel, an instrument powerful enough to cut stone. Thus, the king is also, through the symbols making up his name, associated with elements of strength in the animal and natural world. The same combinations of signs appear twice more on the palette, each time just above the depiction of the main crowned figure. Thus the signs seem to give the king's name. Hieroglyphs consistently appear near the people depicted on the palette, and the writing is quite large in relation to the size of the people, suggesting the importance of the names.

In the main middle section of the back side of the palette, we see the king, wearing the Upper Egypt crown, holding a mace in one hand and grasping the hair of a kneeling captive with the other hand. Beside or behind the king, but on a different level from

| Figure 4.1 |

The two sides of the Narmer Palette, with scenes on both sides commemorating an early king wearing the white crown of Upper Egypt on one side and the red crown of Lower Egypt on the other. The drawings are after Quibell 1898, as presented in Kemp 1989, 42.

him, appears a small figure holding the king's sandals. Facing the king is a falcon. The falcon has been variously interpreted, but all interpretations suggest a strong connection between the falcon and the king, and many suggest a direct relation between the culture's falcon god Horus and the king (Goldwasser 1995, 12). The visual similarity and the spatial positioning of the king and the falcon provide a redundant expression of resemblance. Furthermore, the actions of both appear equivalent; the king tames an enemy, and the falcon restrains a human figure who is depicted with papyrus stalks growing out of his back. Since papyrus commonly grew in the delta region of Egypt, this representation seems to suggest a capture and control of the people of the delta, or Lower Egypt, by the king of Upper Egypt.

Relative size definitely suggests a hierarchical significance here: The king is much bigger than any other humans. The placement of the people below and behind the king, with the defeated enemy at the very bottom, offers an additional component of visual meaning. At the very bottom, two naked enemies under the feet of the king seem to be either floating or running away, looking back at the king.

The Contribution of Words to the Narmer Palette

Earlier palettes and other ceremonial objects offer other types of scenes of battle and hunting, but the representations are generally presented entirely through depictions involving animals. The Narmer Palette shows two new developments: the focus on depicting humans and the addition of pictures that represent names (Woldering 1967, 16). In this early use of writing, the hieroglyphs function to name the players. We do not know what belief system pertained in the particular period in which the Narmer Palette was created, but it is clearly significant that this very early instance of writing uses the capability of recording names of people and places. The names are presented in visual symbols that bring associations of strength and power to the king being named. Although we can't be sure what sounds corresponded to these symbols during this early time period, the symbols often are read now by their relation to later, better-known hieroglyphic symbols. Thus the catfish later corresponded to the consonantal combination *nr;* the chisel later indicated the combination *mr.* Hence the king named by the combination of a catfish and a chisel is now called Narmer. Since Egyptian writing had no signs for vowels, we can only guess at the original vowel sounds in this word (or in any others).

Hypermediacy and Redundancy in the Narmer Palette

In *Remediation,* Bolter and Grusin (1999) move beyond the simplistic but common designation of visual decoding as a holistic process that engages emotions directly (see Seward 1997; Schlain 1998). Instead, Bolter and Grusin distinguish between two approaches to visual media. One style corresponds to the traditional view of visual media in suggesting immediacy and transparency and in presenting a unified visual space. The other style openly presents signs of mediation. Called "hypermediacy" by Bolter and Grusin, this style brings awareness of the media along with some distancing from it. I would suggest that the style of visual media seen here in the Narmer Palette, and in subsequent public multimodal texts in ancient Egypt, is a hypermediated style.

The Narmer Palette shows clearly that the art of the culture that produced it is symbolic, meant to be read. This is not transparent art, pretending to convey reality without mediation. We are not meant to experience the reality of the scene in which a strong king kills an enemy but rather to witness the scene, guided to follow the gaze of the participants and to look at parts in succession; consequently, the viewer is distanced from the whole and led to study and make meaning of each part.

An additional distancing factor is the fact that the two-foot-long palette surface is broken into multiple distinct windows. These windows are placed alongside one another, at times presenting a narrative, at times presenting alternate interpretations of reality. These visual interpretations tend to be repetitive and additive, at times complementary. The narrative line seems to work from the bottom up; the events on the bottom precede events pictured above. There is a loose sense of arrangement by time in the vertical order. But the registers are not arranged in a way that enforces top-down or bottom-up reading. The middle registers dominate, forming the focus of attention. This is a representation of victory, of protection of one's people, and of control of chaos in a violent world, via ritual acts and ordered behaviors. The enigmatic pictures offer a set of meanings for events. Each supplements the interpretation, adding another layer. They seem syncretic more than sequential.

The palette tells us what is important about this early king, or indeed any early king, in the Egyptian system of belief. A king tames enemies. A king brings order, unites the people, and enacts appropriate rituals on their behalf. This work of art makes a theoretical statement about the relationship between kingship, order, and ritual. It simultaneously designs a model spectator, who too is devoted to ritual and to order. Unlike many other cultures' approach to art, this illustrates a conceptual rather than a perceptual way to use the visual medium. Although a juxtaposition of conceptual depictions could be used to create a dialogue between conflicting interpretations, to raise conflicts or questions or doubts, here the multiple windows and frames show a tendency to add, to harmonize, to elaborate. The spatial positioning of the king's palace between the cow goddesses in the top part of the palette places the king within the realm of gods. But the name of the king—involving a chisel and a catfish—along with the depictions of a bull and intertwined panthers communicate the connections of this king with the strongest forces of the natural world. The king is alternately and simultaneously god, the greatest of men, and an impressive power of nature. Each of the multiple meanings is reemphasized in different registers; all coexist in this palette's overall representation of a theory of kingship.

Egyptian Writing Remediates Art

The medium for the Narmer Palette and for other ancient Egyptian monuments is stone, and that material substance bears close relation to the resulting system. The imposing weight and substance and the permanence of stone have often been commented on as significant components in ancient Egyptian usage. Barbara Watterson (1997) points out that the art and writing that arose in Egypt are both particularly suited to chiseling or cutting in stone. The emphasis is on line and contour and on silhouette; the representations are necessarily presented on a plane. The medium of stone restricted scribes and artists in presenting carved details, so they had to find ways to depict things using shapes, with a limited number of lines, to indicate what is meant.

We can see in the Narmer Palette that the Egyptian writing system borrows approaches developed for creating visual representations. The previous medium is, in the terms introduced by Bolter and Grusin (1999), repurposed and remediated. As is the case with the artistic renderings, the hieroglyphs outline representative shapes and units, simply and parsimoniously. William Arnett (1982) studied the predynastic art of Egypt, including the Narmer tablet, by looking for connections between objects represented and later hieroglyphic signs. What he found is quite revealing. Many objects that appear here not necessarily as hieroglyphs, but as renderings of objects, seem very close in form to later hieroglyphic signs for those objects. The Egyptians sought a minimal way of representing an object visually in their art, through simplifying the task of chiseling outlines in stone; these minimal representations then often came to represent the objects iconically as hieroglyphs. The system of writing literally borrowed from the system of visual representation. We can see such borrowing in many objects on this palette, including the falcon, the papyrus stalks, the palace, and the bull.

And as is the case with the art, the hieroglyphs offer down-to-earth and simple formulations that often are somewhat general or ambiguous. A particular sign can be read three ways: pictorially, as representing the entity depicted; phonetically, as representing any words with the particular combination of consonants; and as an ideogram, representing a particular meaning that might relate to the entity depicted (for example, aggression relates to the catfish). Not surprisingly, the phonetic, iconic, and semantic domains can at times yield multiple meanings, an ambiguity as important as the multiplicity of accompanying visual interpretations. Pictorially, the door, birds, and boat above the lines of bodies on the front side of the palette, shown on the right side of fig. 4.1, can be interpreted as a sign for the city where the ceremony took place. That is, the door and boat can indicate a door to the place of boats, or the sea. Together, the signs are taken to indicate the port of Buto (Mark 1998, 96). When read phonetically, the item beneath the bird, above the boat, would mean "great." The bird itself can be read as a swallow, which phonetically would again yield *wr* or "great." The double emphasis then yields "the great port of Buto" (Wilkinson 1994, 94). When read iconically, the bird can be taken to refer to Horus, and the boat to the "sacred bark" of the gods; both signs would then refer to the king as god or to the king's participation in a sacred ritual. Iconically, the signs then yield their meaning differently: Narmer, designated by both the Horus symbol and the sacred bark symbol, conquers the door people (Arnett 1982, 40). Both interpretations constitute valid readings of the set of symbols; both are simultaneously present. In later periods, with more familiar combinations of glyphs, we can be on surer ground with our readings, but some multiplicity, some enigmatic quality, always remains a factor in this language.

Art and writing in this culture emphasized the nonspecific portrayal of individuals. Although the hieroglyphs do individuate the king represented here, the depiction nonetheless simultaneously offers a generic picture of this king. As Wolfhart Westendorf (1968) points out, Egyptian art presents faces as if they are made up of standardizing hieroglyphs (70). Steven Bianchi (1995) notes that Egyptian art presents faces not as signifying identity, but as signifying social class or alien status. The ancient Egyptian culture presented the theoretical ideal, even while naming a particular individual, and later developed the popular textual genres of maxims or instructions, which also tended to present generic advice promoting the culture's ideals. In both writing and art and in works that combined both, the public texts emphasized generic cultural ideals and deemphasized elements specific to individuals.

Visual Language and the Complication of Meaning in Ancient Egypt: The King Scorpion Stela

By the time of the king immediately following Narmer, named Hor-Aha or Menes, we find already a change in the graphic symbol that represented a king as related to the god Horus. Whereas on the Narmer Palette, the Horus falcon was depicted alongside the graphic rendering of the king and the hieroglyphic designation of his name, now the Horus falcon has been moved, placed on top of the palace representation (see figure 4.2). This simplified visual form continued as a way to write the king's Horus name, with a falcon resting on a palace within which appear hieroglyphs representing the name. This new symbolic form appears in several objects from the tombs of this early dynastic era, with different degrees of elaboration of the palace, but all following the same model. Two kings after Hor-Aha, still in the first dynasty of Egyptian pharaohs, we can see that the representation of the king's name now functions simultaneously as both graphic and writing. The King Scorpion stela from Abydos (figure 4.3) also shows that the hieroglyphic representation of the name has been expanded to constitute the entire presentation on a funerary monument.

Although burial chambers in tombs were not accessible to a public, or to family, funerary stelae performed an important cultural and religious function: helping fulfill

| Figure 4.2 |

Redeveloped version of a king's Horus name, here presenting the name of King Hor-Aha, painted on a potsherd. From Clayton 1994, 20.

conditions for a proper afterlife for the individual they commemorated. At no time in the 3,000 years of ancient Egyptian history was it considered automatic that all deceased would move on to, or remain positioned in, a positive afterlife. The beliefs of the culture necessitated the provision of support for a long afterlife, in the form of servants and food, among other things. In very early burials, retainers were killed and buried along with the pharaoh. This practice was abandoned, but models of the necessary workers were then buried with the pharaoh or noble. The need for food was partially addressed in a similar way, by carving and painting food offerings.

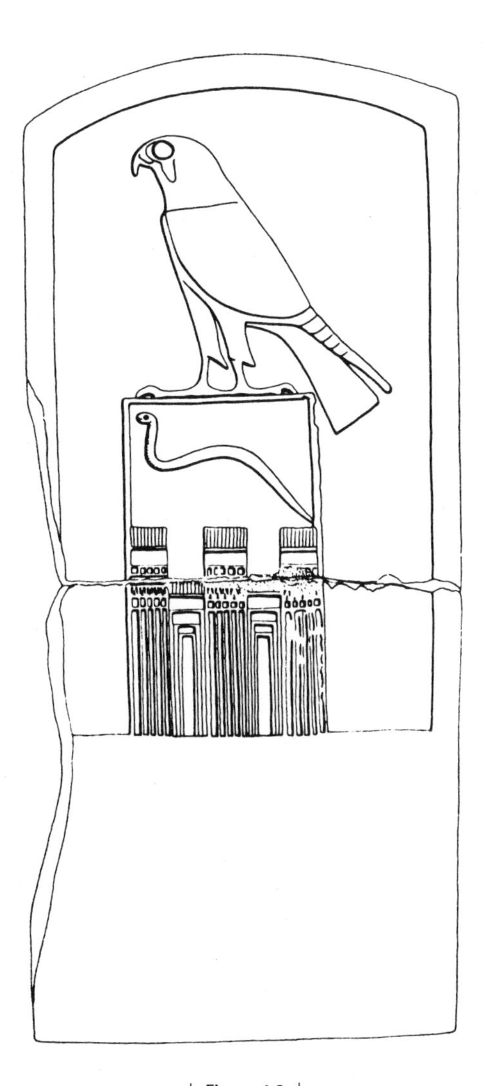

| **Figure 4.3** |

The name of King *Dj* of the first dynasty forms the entire funerary stela for his tomb at Abydos. After Vigneau 1935, 4, as presented in Kemp 1989, 38.

Still, it was believed that food offerings needed to be replenished, and a variety of state and local festivals were developed to form occasions for communities to gather at local burial sites, during which families would feast and celebrate and bring food offerings to their recent dead. The place established for these offerings was an open location in the memorial temple, with at least one stela in place. The stela might be built into a niche in a wall or on a door. It might be outdoors, or it might be in the open in a public room. But it was consistently associated with the offering place, where family and the public could come and perform the offering ritual (Watterson 1997, 79). The burial place could not be entered, but there might be tiny holes in the wall surrounding the burial place to allow the *ba*, the deceased person's spirit in the form of a bird that resided in the tomb, to gain access to the food. Without these food offerings, the individual could not go on existing in the afterlife. The stelae were located where they would be available to families and observers and were accessible to the entire populace (Bunson 1991, 252).

The King Scorpion stela is from the fourth king in the first dynasty of the united kingdom. It dates from approximately 2800 BCE, around 200 years after the Narmer Palette. It was found at Abydos, at a burial site. The entire object is four feet, nine inches high; the part that would have been above ground is about two-and-a-half feet high. The stela is framed with a raised border, within which a falcon sits on a rectangle. In the bottom half of the rectangle, we can see the front and then the view from above a royal palace, with its elaborate recesses. This depiction of a palace as a fortified enclosure, with recessed paneled walls, seen in two parts, became the conventional way of depicting a royal dwelling in the earliest art, from the predynastic era (Arnett 1982). The bird is standing on the palace, though the conventions of later Egyptian art would come to place this bird within the palace (Hornung 1992, 22). The bird is the Horus falcon, as seen on the Narmer Palette. Above the palace, within the rectangle, is a snake. In later times, this sign is a known hieroglyph carrying the sound "dj." This sign thus names the stela as belonging to King *Dj*, known as King Scorpion. Hornung (1995, 1725) suggests that the positioning of the falcon above the palace and the king's name in the palace creates a duality, showing two sides of the king. As king, he resides in the palace. As falcon, he flies to the sky upon his death.

As Tom Hare (1999) points out, the falcon is the largest single figure here; only the palace rivals it in size (81–82). The snake is smaller and subordinate in appearance, but the snake is also at the center: It is the only element shown here other than the falcon and the palace. Since this symbolic representation was repeated in Egyptian art for numerous kings, the new information in this image-text is the snake: the name of this particular king. The palace and the falcon are generic, or given. The snake is posed as spatially related to the falcon, in what Hare calls "a syntax of resemblance" (82). Both are on guard; both look in the same direction. The snake is erect, on alert. So the visual resemblance in these juxtaposed signs tells us that the snake has equivalence to the falcon god; the King Scorpion is the Horus god. The stela can then be taken to read "The Horus Scorpion King," or "The Horus King 'Dj.'"

To "read" this stela, a viewer would have to be familiar with the iconic value of the falcon as the god Horus, with the symbolic representation of the generic framework of a palace, or with both. These icons must have been widely understood and in themselves would yield an understanding that this object belongs to a king, one represented by the scorpion icon.

The Rhetorical Message of the King Scorpion Stela

In *Rhetoric in Popular Culture*, Barry Brummett (1991) focuses attention on the rhetoric of a text by examining in what subject position it places the viewer as that viewer experiences the text (8). Though the King Scorpion stela contains no registers, as does the Narmer Palette, it does create a windowed effect within a raised border. The stela does not address us, but we look into it as into a window from a distance. It literally offers a view inside. It invites one to view icons of the king and the god, but the viewer does not see the king and the god themselves. Despite the increased powers of viewing afforded by the provision of a vantage point from above the place and from the front—a combined viewpoint one could not normally have—the view is from afar, through a window. The viewer is not invited into the palace, to see the king on the throne, the king as a person. What the king looks like is not what is considered important. Instead, his role in the scheme of things is what matters. The image-text is structured to keep the viewer at a distance, offering a window on an ordered, harmonious, hierarchical world of roles and relationships.

With minimal graphics and text, this early representation thus offers a sense of the values of the culture, a sense of how to live within the culture. The stela embodies a strong epideictic message, conveying the primacy of the eternal realm of the gods and of the king as its representative and embodiment. It tells a viewer how to live one's life, how to understand one's relationship to one's king and to the king's function in maintaining the well-being of the kingdom according to the principles of the divine realm. The immediate context of this stela—its *kairos*—is the need to provide a permanent form of the name of the king for the afterlife and a place of offering. But the stela also speaks of *kairos* to the viewer, conveying a sense of what the viewer should take up as contexts to respond to in action. It emphasizes the centrality of the gods in preserving the kingdom from chaos and the centrality of the king in doing so on Earth; what is first and foremost is that the falcon god is in/on the palace.

As Hare (1999) points out, the elements of the king's name, although constituting a written object, cannot leave behind their graphic existence when they appear in such an artistic representation. The snake is a phonetic representation of a name, but it is also a snake, a feared part of the natural world. The falcon stands for Horus, but it is also a falcon resting on top of a building. The palace tells us this is a name of a king, but the palace also represents a particular building, inhabited by this king, unavailable to others. The very graphic symbols that so effectively indicate the godly and earthly status of the king, embedded in the written form of his name, clash with the reality of this object's function in its setting. For this is a funerary monument found in a cemetery at Abydos. Once this stela is erected, the Scorpion King is no longer alive, no longer living in his palace, no longer the living embodiment of the Horus god. Although the stela offers a statement of the mythic centrality of the king, at the same time it presents a statement of the contradictory nature of reality: The king is a god figure, but the king is mortal. The contradiction is contained within an ordered world; the stela as a whole is particularly spare and tidy, elegant and serene. The graphic nature of this system of writing creates a tension that remains in all Egyptian funerary objects. The contradictory outlook is fundamental to the Egyptian understanding of death, an eternal existence, and to the rituals surrounding death. The mode of writing in its graphic base fosters the presentation of coexisting dissonant meanings.

The Significance of Framing in the Making of Meaning: Iykhernofret's Stela

For several hundred years following the Scorpion Stela, not much written text appeared on the funerary stelae (Hornung 1992). But continuous text did eventually begin to appear on stelae and tomb inscriptions for members of the bureaucracy. By the twelfth dynasty, the stelae were covered with lines of text. At times, the visual element seems to offer a visual representation of the textual passage provided, reinforcing the meaning. Such is the case in numerous stelae that show an image of the individual for whom the stela is erected alongside some text identifying the individual. Such an image may depict the individual in a gesture of worship, replicating the purpose of the text as an offering to a god. In these cases, the graphic depicts the gesture that the text enacts. Often, the texts of Middle Kingdom stelae offer justification for receiving offerings, justification for an afterlife. Such a justification may appear in the first person, yet it still reads as a generic list, showing that the individual followed the culturally sanctioned values. The images too show an ideal, not a particular individual. The text and images often face one another, depicting their equivalence.

In another approach common throughout the Middle Kingdom, justification for the afterlife is presented via a close relationship to the king: one of dutiful service. Such is the case in a stela belonging to an official named Iykhernofret, who had been sent by King Senusert III to take responsibility for the ritual functions dedicated to the god Osiris at Abydos. Iykhernofret's stela contains a large quantity of text, which communicates his close connection to the king, who charged him with a high degree of responsibility. Figure 4.4 presents the dense appearance of the stela, and figure 4.5 offers a simplified schematic illustrating what appears in each of its major parts.

The textual portions in the center of this stela consist of two parts. One copies the decree by which the king announced Iykhernofret's new position and its charge. This is the equivalent of our copying a letter announcing a new job and its responsibilities on a gravestone, except that the text expresses the king's having taken an adult Iykhernofret into the palace in personal terms of his having gained a foster son, a new family member. The second portion of the textual section offers Iykhernofret's recital of what he did in the position to which he had been assigned, in fulfillment of the king's charge. This is not a list of virtues, but a list of duties performed, perhaps the equivalent of a resume today. The text thus suggests a careerist ideology: Serve your king well, and you will do well.

The two textual portions in this stela are framed within multiple windows that both reinforce and dialogue with this careerist emphasis. The framing consists of three parts: an outer border, with text inside; a semicircular area below this border and above the two registers of long, continuous text; and a lower portion with its depiction of Iykhernofret receiving offerings from what are labeled as relatives. Each portion interacts with the text and with the other portions.

The outer border offers a space within which a ladder of names climb to the summit, where we find a graphical depiction of the falcon god Horus, with wings outstretched, seen as if we are level with the bird as it flies. The text within the border is itself framed between graphical depictions: of Iykhernofret at the bottom on each side, and of the Horus god at the top. Iykhernofret's name and titles are given just above each small human figure; in fact, the human figure in each case serves a grammatical function as a component of the text: as a determinative, to indicate that the text names a man. The image is thus a crucial part of the text. Above Iykhernofret's names and titles on each side appear the titles of King Senusert, and above that the titles of Horus. For Iykhernofret, a nonroyal official, to appear within the same border frame as the king

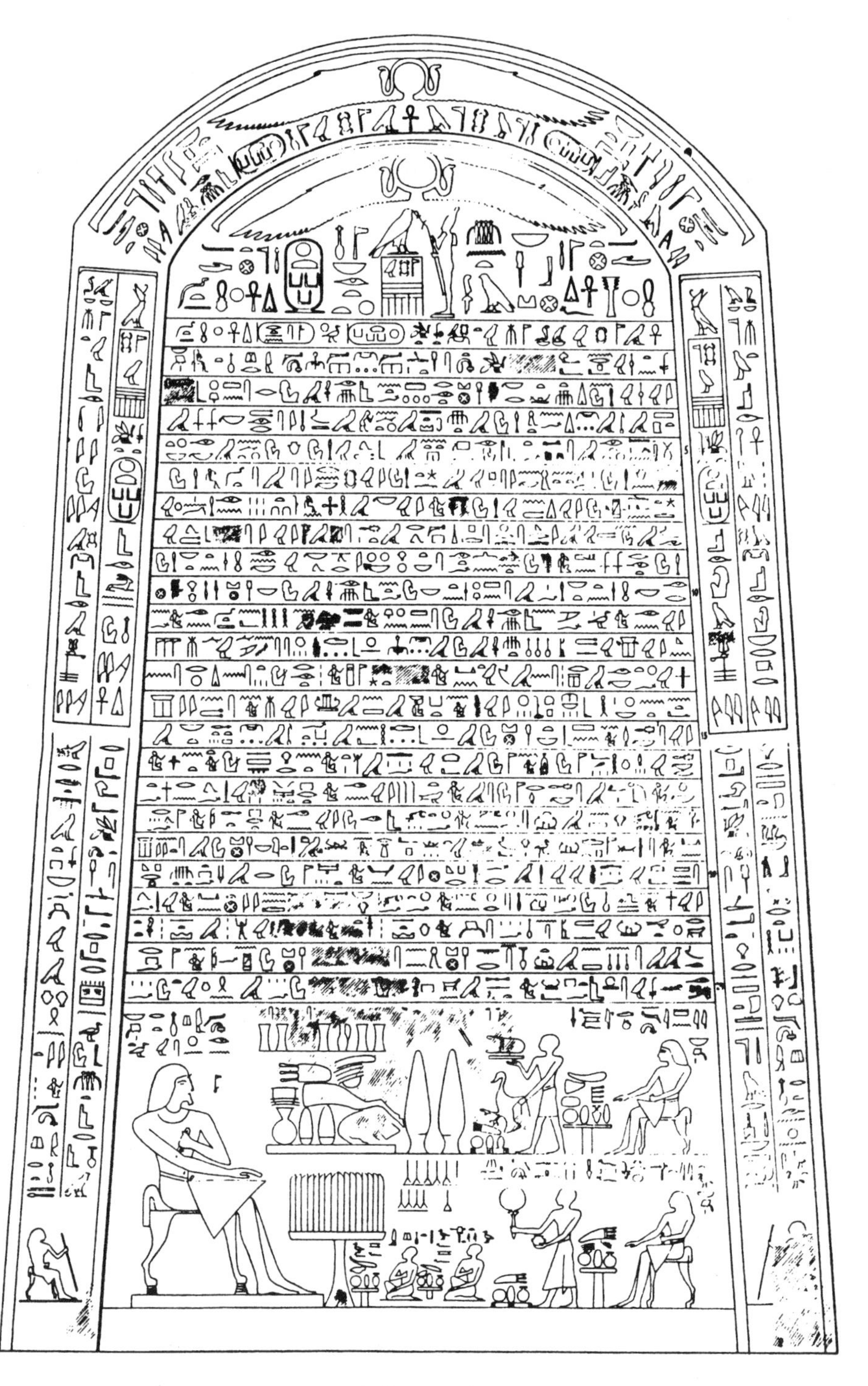

| Figure 4.4 |

Stela of Iykhernofret, a twelfth-dynasty official under King Senusert III. The stela was found at Abydos, associated with the cult of Osiris. Line drawing after Shafer pl. I, as presented in Hare 1999, 36.

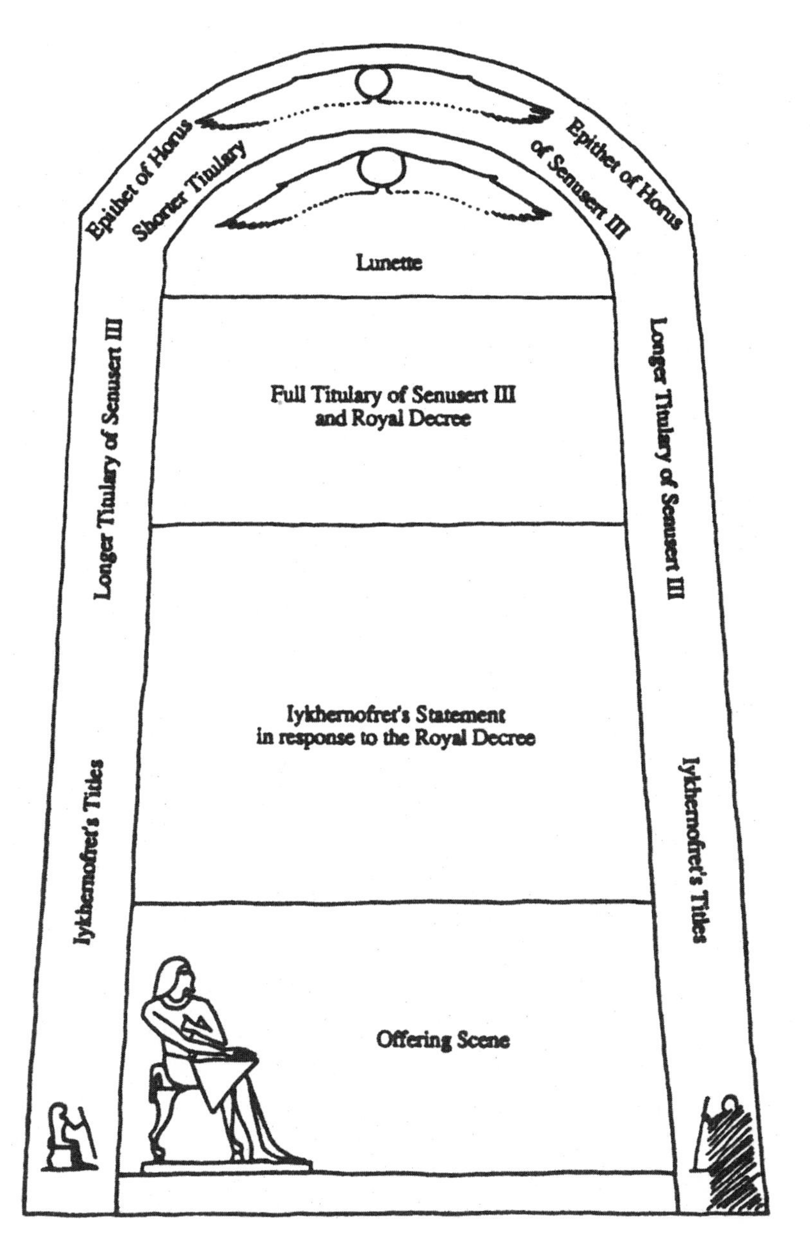

| Figure 4.5 |

Schematic drawing of the stela of Iykhernofret, showing the major spatial divisions on the stela. As presented in Hare 1999, 38.

and the Horus god is itself a visual device with powerful symbolic meaning. Iykhernofret, though at the bottom of this hierarchy, infinitely smaller than the Horus god, shares the space of the exalted cultural powers. He is shown supporting them from a lower rung of power in comparison with their status, but still on the same ladder.

The semicircular window below the border provides a second frame for the text; it repeats the graphic figure of the Horus falcon, emphasizing through the redundancy that everything here functions within the protective sphere of the Horus god, for whom the king is a living incarnation. Yet this stela is erected at Abydos, a center for the worship of the god Osiris. Under the second representation of the Horus figure, a small Osiris figure is seen giving life to the king, by touching an ankh symbol to the Horus name of the king. The entire king-oriented text in the middle of the stela functions visually under and within the life-giving and life-protecting functions of the culture's many gods, and literally under the protective wings of Horus.

As Hare (1999) points out, the Horus falcon with outstretched wings, or the depiction of some other god in its place, was generic in funerary stelae and so might be taken by viewers as not offering new or central meaning but functioning as a background. The repeated Horus figure offers a variation of the generic, however, doubling the emphasis, potentially calling more attention to the generic portrayal of the Horus falcon. In this visual framing device, the king in the center of the stela functions to carry out on Earth the will of the gods in preserving order and keeping chaos in control. And Iykhernofret supports the king in a world dedicated to the gods.

Other Middle Kingdom or later stelae frame their contents in other ways. Instead of words in a border, decorations might appear in the border or above the text. The decorative elements are never random, never chaotic, and never without meaning. The decoration often offers a regular pattern repeated, whether consisting of lines or abstract depictions of plants or other design elements. In all such cases, the framework presents a border that depicts an ordered realm. For the ancient Egyptians, life was centered on the need to keep disorder at bay, under control. The propensity to use repeated lines or patterns, even to divide a surface into an ordered arrangement of rectangular units, reinforces and reflects the value placed on order.

A Syntax of Difference: Multimodal Text and Multiple Points of View

The bottom of the Iykhernofret stela shows Iykhernofret receiving food offerings, in a standard scene, commonly represented in varied forms on funerary stelae. The small figures bringing the offerings are listed as Iykhernofret's family. They are tiny, almost faceless. The text in that section repeats the standard offering formula that appears regularly on stelae; in this formula, the king makes an offering to the gods so that the gods will provide offerings for the king to give to the deceased, thus ensuring the deceased's afterlife. Yet at the same time, the family members are shown actually carrying out this duty of providing offerings. The formula suggests that the king will bring the offering, yet the king would never literally do so because the chain of obligation, some of it ritual in form, will provide the necessary elements. The king's participation is needed in communicating with the gods, in serving as intermediary between the family members and the gods. The gods are ultimately in control, but the king is required as their representative on Earth. The formula does not present the king as autonomously powerful, but as subject to the will of the gods, as having to make offerings to the gods and to fulfill rituals to the gods, in order to ensure a good life on Earth and in the afterlife for his people.

The relative sizing of the graphics speaks powerfully of the relative positions of other members of the hierarchy in relation to the gods of the culture. The family members appear as the least significant elements here, yet they are the ones who actually provide for Iykhernofret's continuation in the afterlife. Although this stela offers an epideictic message about what is important in this culture—serving the king, who himself has been created by and is protected by the gods—it also offers something of a counterstatement about the importance of one's family obligations. The text presents Iykhernofret as having acquired a new family: a father, Senusert; a father, Osiris. But the visual elements of the text clearly convey a hierarchical set of connected obligations. Iykhernofret's own family, which is absent from the central textual portions, is presented visually and by name as having a major role in the chain of responsibility. Unfortunately, however, Iykhernofret's responsibility and devotion to them seems entirely absent in the text. In the end, this stela rests on a point of tension in its guidelines for living one's life and its understanding of provision for the afterlife. The dissonance is unresolved; both points of view stand.

Presenting the Unpresentable: The Case of Hatshepsut and the Visual Rhetoric of Accommodation

All the examples up to this point show a conformance with the values of the elite cultural tradition. Yet when a pharaoh or other member of the elite was in a position of doing something outside of the traditional norm, that individual could turn to the same principles of using image-texts to position the deviance within the standard ideology and iconography. The system could in fact accommodate fairly major deviations, by depicting their connections to and grounding in authoritative ideology and myth. In the eighteenth dynasty, a queen named Hatshepsut took power as pharaoh when she was coregent with a very young brother, Tutmothis III. Hatshepsut became a pharaoh, but female pharaohs were not customary. The graphic portrayals of Hatshepsut as pharaoh all show her as male, with a male physique; once she became pharaoh, even portrayals of Hatshepsut as a baby showed her as a male. Whereas in the early period, the gods were depicted as animals, by the New Kingdom, fifteen hundred years later, they are presented in human form, with distinctive headdresses to identify them as particular gods. In human form, they can speak, and their text can provide sanction for deviance in a variety of ways. In Egypt, women were excluded from these representations of kingship in embodied form; for the visual record, the female pharaoh portrays herself in a male body.

Typically, the public texts, on obelisks and on walls, justify Hatshepsut's kingship by portraying her as sanctioned by a god, as having divine lineage, even as having been declared as successor by her father. Since her father did not declare her as future king, she is taking liberties in rewriting history to establish a harmonious succession, sanctioned by both the previous pharaoh and the gods (Tyldesley 1996). Figure 4.6 shows a typical approach to self-justification taken by Hatshepsut. This is an upper portion of an obelisk, no longer standing, from the god Amun's temple at Karnak. Here we can see a more elaborate extension of the device seen in the Narmer tablet, where the Horus falcon image is presented alongside the image of the king with the enemy figure. Now the god Amun is seen as a human figure with the capability of speech. The suggestiveness of the early combination of text and art has become more explicit, with the addition of a textual passage by the god that acknowledges Hatshepsut's divine lineage as daughter of Amun and her rightful place as pharaoh, while also bestowing protection and life. For those who could not read the text, the image of Amun reaching out in a protective embrace to Hatshepsut would convey the meaning. Amun's arms, which seem to assume a very unnatural position, in fact form a hieroglyph: *ka*, the force that creates and preserves life (Westendorf 1968, 96, 101). Hatshepsut kneels in front of Amun, in a depiction that symbolizes his responsibility for giving her life. In this portrayal, as is typical, the adult Hatshepsut wears male clothes and appears to have no breasts.

In Hatshepsut's mortuary temple, on the other side of the Nile facing Amun's temple, the walls contain a sequence of large-scale murals depicting her divine lineage. A number of the depictions focus on her lineage as daughter of the god Amun; one presents her father's declaration of her succession. One scene shows another god, Khnum, and his wife Hecket, having created the baby Hatshepsut and giving her life. The baby is portrayed as a male, with male genitals and a male hairstyle. So the image-texts offer a multiplicity of sanctioned ways of interpreting reality. For the ancient

| **Figure 4.6** |

Portion of broken tip of obelisk erected for Hatshepsut at Karnak, showing her as a male figure, receiving life from the god Amun. The text, in the words of Amun, presents her as his daughter. From Clayton 1994, 105.

Egyptians, each god was associated with a separate story of creation; all could coexist with equal force, despite their differences.

Notably, whereas the visuals all portray Hatshepsut as male, the texts always present her in the feminine forms. Only the elite could read the text, but all societal groups could make some sense of the images. Those images here present one version of reality: Hatshepsut as male. The hieroglyphic texts, which could be read only by the elite, consistently use the feminine pronoun forms to refer to Hatshepsut in all image-texts. Thus the text and the images could tidily tell different stories. The combination of media here allows for differential communication for different audiences. The juxtaposition accommodates both stories as true: The pharaoh's gender is female, but the pharaoh in this society is a male role, so the occupier is male, even though she is female. The elite would have known that Hatshepsut was not a male; the commoners would not. She couldn't have ruled in peace for over fifteen years without the support of the elite. But the façade of the male pharaoh, the cultural ideal, could also be maintained for those who had no knowledge otherwise.

The Visual Language of Ancient Egypt

We have seen that the later image-texts of ancient Egypt carry on the tradition established in the early versions. The differences in multiple meanings are allowed to stand simultaneously, to coexist in a complex presentation. At times, the text and the images tell different stories, as in the representations of Hatshepsut. Whereas Greek rhetoric developed invention techniques to find points of dissension in a dialogue, with an elaborate system of stasis questions to determine the differences and the points at issue, Egyptian multimodal rhetoric is a rhetoric of accommodation to the ideal. And the ideal could encompass contradictory elements. Here the invention process seems focused on seeking points of connection, points of resonance with the ideal framework. The image-texts involve a weaving of authoritative voices and ideal cultural topoi; they name and interpret reality in culturally sanctioned ways.

In the Hatshepsut image-texts, the images tell a different story than does the text. In the Scorpion stela and the Iykhernofret stela, the divisions are not so simple. In the Scorpion stela, the graphic elements contain internal contradictions. In the Iykhenofret stela, the different textual sections present contractions, particularly in the text naming the family, which contradicts the offering formula and the long section suggesting that service to the king is primary for an afterlife. In this stela, the graphic elements present yet another view, involving the primacy of the gods and the place of the king and his servants in relation to the gods. Although the different emphases of the parts are subsumed by the overall emphasis of the framing device, the internal contradictions nevertheless remain.

These examples all illustrate how the culture of ancient Egypt followed the same principles that had been established for the creation of visual media to treat writing in public texts from the very earliest stages over a period of about fifteen hundred years. Both text and graphics are presented in an accumulation of windows, generally rectangular. Both operate syncretically, providing a multiplicity of meaning. The forms of both graphic figures and written letters or words involve outlines of recognizable objects, presented simply. Both graphic representations and written texts are built on similar principles of generality and universality, offering ideals rather than particular individuals or situations. Both are ideologically based, presenting and reinforcing the ideals and values of the elite culture. They operate in conjunction, in light of one another. The graphics cannot really be divided from the pictures: the text is art, and the graphics are text, at times literally, at times figuratively. From very early on, virtually no public texts presented graphics without some written text, even if that text provided only a name. The meaning of any public image-texts in this culture depended heavily on the visual media, even when the written text dominated. Betsy Bryan (1996) suggests that the two media speak to different audiences, one literate, one not. I would agree that such is the case at times. But I would contend that the rhetorical strategies in the early literate culture's practices of multimodal presentation are more varied and more complicated. Erik Hornung (1995) points out that the ancient Egyptians possessed a highly sophisticated sense of the possibilities in using visuals, with twenty or more words for different kinds and uses of pictures (1729). They developed a visual language that included text as graphically meaningful and included graphics as textually meaningful. The rhetorical use of multiple media in conjunction both enabled and enacted the presentation of complex statements in varying ways, often based on a multiplicity of coexisting meanings.

This examination of the multimodal rhetoric of ancient Egypt also shows that the principles developed by Gestalt psychologists early in the twentieth century, explaining a range of ways that humans interpret spatial elements, can be seen at work in the conventions of ancient Egypt for presenting meaning in visual representations. That is, the Gestalt principles, as listed by Horn (1998), suggest that meaningful visual attributes involve proximity, similarity, enclosure within common regions, connectedness, alignment or directional continuation, and closure. Horn uses these principles as "the foundation of visual language syntax" (75). The analysis here has attempted to show that these attributes prove helpful in explaining the visual meaning of the public texts of ancient Egypt and the relation of language and text in those image-texts. The ancient Egyptians 5,000 years ago seem to have built their visual and written language on such fundamental principles, embedded within the specific context of a particular culture as it underwent changes. In ancient Egypt the visual language provided the capacity for presenting complex coexisting meanings within compact and accessible image-texts. The multimedia texts of ancient Egypt exist as collages, displaying a range of components that mediate one another, frame one another, and dialogue with one another. The system of combining media offered culturally sanctioned ways to accommodate differing and alternate understandings as simultaneously valid, without the need for erasure or defeat of the differences.

Some Implications for the Digital Era

In today's digital media, we are seeing a repurposing of a variety of earlier media. Over the past ten years, I have been teaching courses analyzing new types of multimedia electronic texts, mostly on CD-ROM, but some on the Web. These texts remind me of the early stages of moviemaking. Early movies transported a then-familiar genre, the theater, and only gradually began to exploit the new possibilities afforded by the new medium. At first, we had plays transposed to the screen. Now we have multimedia works arising from one or another of the various media. For example, a multimedia version of Nicole Brossard's feminist novel *Mauve Desert* arose from the video or film world. Adriene Jenik (1997) created much more than just a movie or video version of the book; instead, her interactive multimedia text found ways to interactively illustrate, read aloud, or show the actual text, at times to enact it, at times to comment on it. The 1996 winner of the International Multimedia Award (the Milia D'Or Grand Prize) was *Scrutiny in the Great Round*, and it came out of the art world, adding music and text and interactivity. Tennessee Rice Dixon, Jim Gasperini, and Charlie Morrow created in *Scrutiny* an artistic experience, quite soothing, beautiful, and romantic. A 1995 CD by Elliot Earls, *Throwing Apples at the Sun*, coming from the field of visual design, creates a postmodern, disjunctive experience, intriguing and frustrating. Anne Wysocki (this volume) points out the ways these multimedia CDs arising from the art and design fields do not fulfill the predictions made by those scholars in textual fields. One 1995 cult multimedia CD by Laurie Anderson came out of the avant-garde music and performance art worlds; called *Puppet Motel*, it was based as well on the paradigm of video games. The textual world has generated hypertextual literary works with graphics, such as Michael Joyce's (1996) *Twelve Blue*, or Ed Falco's (1995) *A Dream with Demons*. We are also seeing works of electronic literature by people with backgrounds in both art and literature, such as Shelley Jackson (1995), who produced *Patchwork Girl* for

Eastgate, and Helen Cho, whose 1996 Web work entitled *Quiet Foxes* won recognition in that year in the New Voices/New Visions multimedia competition. Christy Sheffield Sanford's (1996) illustrated Web work *Safara in the Beginning* also won early recognition in the New Voices/New Visions competition. The television world spawned a series of interactive online Web soap operas, as well as documentary works such as one on the Zapatista struggle recognized in a New Voices/New Visions competition.

Each of these early digital multimedia texts builds on the paradigms and processes of the media field from which it arises, adding in new media to complement and expand the range. The integration of media promises to present complex, textured thought without reducing that thought to uncomfortably univocal textual argument. Not all approaches to multimedia—to the combination of images and text—would necessarily allow for the multiplicity of meaning that I advocate. Disciplinary backgrounds lead practitioners to envision and enact different possibilities and goals when combining media. Early digital multimedia texts were created to realize extensions of goals embedded in the fields from which they arose, and these goals will inevitably differ as different disciplinary cultures explore and develop the possibilities of integrating media in new combinations. It seems promising that a significant number of the pioneers in the early creation of multimedia works in the 1990s were women. I would hope that the wide variety of approaches arising in the different fields would maintain the possibility for rich and complex meanings, for alternative rhetorics. According to Horn (2000), the fragmentation and complexity of contemporary life demands a mode of creating meaning that can convey the depth and detail and complexity of our world, and visual language offers that opportunity. This is an opportunity on which we need to act, to ensure its viability.

Works Cited

Anderson, Laurie. 1995. *Puppet Motel.* CD-ROM. New York: Voyager.

Arnett, William S. 1982. *The Predynastic Origin of Egyptian Hieroglyphs.* Washington, D.C.: University Press of America.

Bianchi, Steven. 1995. "Ancient Egyptian Reliefs, Statuary, and Monumentary Paintings." In *Civilizations of the Ancient Near East,* ed. Jack Sasson, 2533–2554. New York: Charles Scribner's.

Bolter, Jay David and Richard Grusin. 1999. *Remediation.* Cambridge, Mass.: MIT Press.

Bryan, Betsy. 1996. "The Disjunction of Text and Image in Egyptian Art." In *Studies in Honor of William Kelly Simpson,* ed. Peter der Manuelian, 161–168. Boston: Museum of Fine Arts.

Brummett, Barry. 1991. *Rhetoric in Popular Culture.* New York: St. Martin's.

Bunson, Margaret. 1991. *The Encyclopedia of Ancient Egypt.* New York: Facts on File.

Cho, Helen. 1996. *Quiet Foxes.* Hypertext. Available: <http://www.quiet-time.com/>.

Clayton, Peter. 1994. *Chronicle of the Pharaohs.* London: Thames and Hudson.

Dixon, Tennessee Rice, Jim Gasparini, and Charlie Morrow. 1995. *Scrutiny in the Great Round.* CD-ROM. Santa Monica, CA: Calliope Media.

Earls, Elliot. 1995. *Throwing Apples at the Sun.* CD-ROM. Greenwich, CT: The Apollo Project.

Falco, Edward. 1995. *A Dream with Demons.* Disk. Watertown, MA: Eastgate Systems.

Goldwasser, Orly. 1995. *From Icon to Metaphor: Studies in the Semiotics of the Hieroglyphs.* Fribourg, Switzerland: University Press Fribourg.

Hare, Tom. 1999. *ReMembering Osiris.* Stanford: Stanford University Press.

Horn, Robert E. 1998. *Visual Language: Global Communication for the 21st Century.* Bainbridge Island, WA: MacroVu.

Horn, Robert E. 2000. "The Representation of Meaning: Visual Information Design as a Practical and Fine Art." Speech presented at the InfoArcadia Exhibit, The Stroom Center for the Visual Arts, The Hague, Netherlands, April 3. Available: <http://www.stanford.edu/~rhorn/VLbkSpeechMuralsTheHague.html> (accessed December 2001).

Hornung, Erik. 1992. *Idea into Image.* Princeton: Princeton University Press.

Hornung, Erik. 1995. "Ancient Egyptian Religious Iconography." In *Civilizations of the Ancient Near East,* ed. Jack Sasson, 1711–1730. New York: Charles Scribner's.

Jackson, Shelley. 1995. *Patchwork Girl.* CD-ROM. Watertown, MA: Eastgate Systems.

Jenik, Adriene. 1997. *Mauve Desert.* CD-ROM. Los Angeles: Shifting Horizons Productions.

Joyce, Michael. 1996. *Twelve Blue.* Available: <http://www.eastgate.com/TwelveBlue/> (accessed March 1997).

Kemp, Barry. 1989. *Anatomy of a Civilization.* London: Routledge.

Kress, Gunther. 1998. "Visual and Verbal Modes of Representation in Electronically Mediated Communication: The Potentials of New Forms of Text." In *Page to Screen: Taking Literacy into the Electronic Era,* ed. Ilana Snyder, 53–80. London: Routledge.

Kress, Gunther, and Theo van Leeuwen. 1996. *Reading Images: The Grammar of Visual Design.* London: Routledge.

Mark, Samuel. 1998. *From Egypt to Mesopotamia.* London: Chatham.

Mitchell, W. J. T. 1994. *Picture Theory.* Chicago: University of Chicago Press.

NCTE Resolution. 1996. "On Viewing and Visually Representing as Forms of Literacy." Available: <http://www.ncte.org./resolutions/visually96199.shtml>.

New Voices, New Visions. 1995. CD-ROM. New York: Voyager.

Parkinson, Richard B. 1991. *Voices from Ancient Egypt.* Norman: University of Oklahoma Press.

Perniola, Mario. 1995. *Enigmas: The Egyptian Moment in Society and Art*, trans. Christopher Wodall. London: Verso.

Quibell, J. E. 1898. "Slate Palette from Hieraconpolis." *Zeitschrift fur Agyptische Sprache* 36.

Sanford, Christy Sheffield. 1996. *Safara in the Beginning: A Moving Book.* Web novel. Available: <http://web.purplefrog.com/~christysafara/safara.html> (accessed March 1997).

Schlain, Leonard. 1998. *The Alphabet versus the Goddess.* New York: Viking.

Seward, Anne Marie Barry. 1997. *Visual Intelligence.* Albany: State University of New York Press.

Silverman, David. 1991. "Divinity and Deities in Ancient Egypt." In *Religion in Ancient Egypt*, ed. Byron Shafer, 7–87. Ithaca: Cornell University Press.

Spencer, A. J. 1993. *Early Egypt.* Norman: University of Oklahoma Press.

Stafford, Barbara. 1996. *Good Looking: Essays on the Virtue of Images.* Cambridge: MIT Press.

Tyldesley, Joyce. 1996. *Hatchepsut: The Female Pharaoh.* London: Viking.

Vigneau, A. 1935. *Encyclopedie photographique de l'art: Les antiquities egyptiennes du Musee du Louvre.* Paris.

Watterson, Barbara. 1997. *The Egyptians.* The Peoples of Africa Series. Oxford: Blackwell.

Westendorf, Wolfhart. 1968. *Painting, Sculpture, and Architecture of Ancient Egypt*, trans. Leonard Mins. New York: Harry N. Abrams.

Wilkinson, Richard. 1994. *Symbol and Magic in Egyptian Art.* London: Thames and Hudson.

Woldering, Irmgard. 1967. *Gods, Men and Pharaohs: The Glory of Egyptian Art.* New York: Harry N. Abrams.

| 5 |
DIGITAL IMAGES AND CLASSICAL PERSUASION

Kevin LaGrandeur

> The digitization of images inevitably strips away their context and allows the machine, or rather its programmer, to define new contexts.
>
> —Jay David Bolter, *Writing Space*

Introduction

The abilities to digitize and contextualize images on the computer required, through the late 1980s, some degree of mathematical expertise. Digital graphics are really pictures made by equations and were originally constructed piece by piece. Now, however, the ease of digitizing photographs and drawings has made the Web's graphic landscape much more accessible to the average person. Thus, the statement by Bolter that begins the chapter now has added implications. Where once only words were malleable enough to be widely wielded as a rhetorical tool, in the latter half of the 1990s the digital image became prevalent, easy to manipulate, and consequently, easy to recontextualize, meaning that now just about any image is available to any computer user for any occasion. To use Bolter's terminology, the "interpenetration" of textual and pictorial space in digital environments, especially the World Wide Web, has increased markedly, so that the predominance of the digital image now rivals that of the digital word. Indeed, a number of thinkers have noted the digital image's ascendancy in communicating information via the computer.[1] But how are we to think about, to analyze the rhetorical dimensions of these images? Both static and moving images can be intensely affective, of course, as print, film, and television have taught us; but what model can we use to assess the persuasive impact of the image in the realm of information technology—specifically, in environments like the Web, a realm where there is an interdependence between text and graphics, as well as an interactivity between reader and writer/programmer/rhetor?

Many have turned to postmodernism to theorize the digital medium in general. The gist of such theorization is that the characteristics of new media like the Web—collage, hypertextuality, multimodality, and nonlinearity, for instance—enact the postmodern *texte*. The focus of this thinking tends to be on aspects of chaos and fragmentation represented by such digital media. But one can also approach these media from another viewpoint, focusing on them as integrative, intertextual, and complex. Notable among those who have approached digital media from this angle are Gunther Kress, Jay David Bolter, Richard Lanham, and Kathleen Welch. The latter two authors, though they sometimes make use of postmodern theory, have successfully used classical rhetoric as their foundation for analyzing computer media. Lanham (1993) discussed digital textuality, including some focus on the digital image, in these terms back in *The Electronic Word*. More recently, Welch (1999) has explored how Isocratic rhetoric may provide a way to think about modern video-based communication, a category in which she includes computers. This chapter owes a debt to these others, and proceeds in their spirit, but focuses particularly on using classical rhetoric as a way of thinking about the persuasive power of computer-based images.

Why Refer to Classical Rhetoric?

There are good reasons for looking at the digital image in classical terms. In a general sense, as Lanham (1993) contends, few models provide a "frame wide enough" to explain the "extraordinary convergence of twentieth-century thinking with the digital means that now give it expression"; therefore, he continues, "to explain reading and writing on computers, we need to go back to the original Western thinking about reading and writing—the rhetorical paideia that provided the backbone of Western education for 2,000 years" (51). Because, with increasing bandwidth, images have become ever more integral to the computer-based reading and writing process since Lanham wrote this passage, I would argue that what he says applies to images, as well. Moreover, as Welch (1999) puts it, classical rhetoric is pertinent to the new communication technologies because "classical Greek rhetoric" is "intersubjective, performative, and a merger of oralism and literacy" (12), and these qualities are common to the technologies in question. I would add to her assertion that these qualities are especially common to the realm of Web-based presentation. For instance, as I shall discuss later, images on Web sites act as part of an argument by parataxis, which, as Eric Havelock has maintained, is characteristic of oral rhetoric, the heart of the classical system (see Lanham 1991, 108). Finally, there is good reason to redeploy classical rhetoric to examine the persuasive value of digital images because, as I intend to show by presenting the thoughts of some of its most notable thinkers, classical notions provide us with excellent, codified ways to think about the persuasive efficacy of images and words as interdependent and interactive things.

The Image and Classical Rhetoric

It might be extremely difficult to have a true argument, with the give and take that "argument" implies, using *only* visual images. Yet the potential of the image to move its viewers was recognized by ancient rhetoricians, and thus a correlation between it and verbal imagery has been an important component of persuasion since classical times. The theoretical basis for seeing images as modes of persuasion lies in Aristotelian rhetoric, which stipulates that the speaker's ability to arouse emotion in his audience and his ability to cultivate an impression of credibility with them are, in addition to evidence and logic, extremely important persuasive elements.

In practical terms, the precedents for the use of images and imagery to instill emotion or credibility can be found in two slightly different classical traditions. One tradition, stemming from Aristotle and continuing with the early Greek orator Gorgias, concerns the affective similarity of images and words: In his *Encomium of Helen*, Gorgias equates the emotive power of the image with that of persuasive speech. The other tradition, most famously associated with the Roman writer Horace, emphasizes how the poetic image can be persuasive: In discussing poetry's instructional potential, Horace mentions the similarity of poetry to pictures. This Horatian idea became very popular among literary critics and rhetoricians, especially those of the Neoclassical era. In fact, as is exemplified in the theories of the eighteenth-century rhetorician George Campbell, these slightly different traditions of Gorgias and Horace appear to have mingled together over time, so that poetry, visual images, and persuasive speech and composition became interdependent. In the age of the pixelated image, which has given rise to everything from television advertisements to hypermedia, the rhetorical principles codified by Aristotle are still important: Fluency with images and their use has become crucial to controlling credibility and creating emotional appeal, and even, to some extent, logical appeal.

The Aristotelian Basis for Linking Images and Persuasion

One reason Aristotelian rhetoric provides a good basis for discussing the image as a persuasive tool is that Aristotle's definition of rhetoric is broad enough to encourage it: He defines rhetoric as the art of finding "in any given case the available means of persuasion" (*Rhetoric* [1984b], bk. I, chap. 2.). In the years since he wrote those words, our "available means" have expanded considerably, especially with the advent of electronic gadgetry like the computer, which has evolved from a solely mathematical tool into, as Richard Lanham (1993) has proclaimed, a rhetorical medium.[2] In this respect, Aristotle's definition of the different means of proof available to the orator provides the most important means of discussing the persuasive image. Besides nonartistic proofs, or what we would call "hard evidence," Aristotle specifies that three artistic forms of proof are also important to argument: *logos* (an appeal to reason), *pathos* (an appeal to the emotions), and *ethos* (the appeal implicit in the speaker's character and credibility). Although he felt that only "the bare facts" *should* be weighed in any kind of decision and that a plain rhetorical style should suffice, Aristotle grudgingly conceded that because of the "defects of the hearers" such things as artistic appeals are necessary (bk. III, chap.1). Accordingly, he devotes much of his discussion in the *Rhetoric* to examining artistic forms of appeal, their delivery, and how they may affect the psychology of a given audience. In this context, he discusses the effects of images rendered in words. "Prose writers must," he says, "pay especially careful attention to metaphor," for it gives writing a "clearness, charm, and distinction as nothing else can" (bk. III, chap. 2).

The source of the image's power is clearly in its emotional appeal. As Aristotle points out in his *Poetics*, "Though the objects themselves may be painful to see, we delight to view the most realistic representations of them in art, the forms for example of the lowest animals and of dead bodies" (1984a, chap. 4). The source of the image's emotional appeal clearly depends on its mimetic quality; but it also depends significantly on the artist's manipulation of the image, on its rendering, not *just* on its "realistic" imitation of nature. This is why one may have no familiarity at all with the original object rendered in an image and still experience emotion when viewing the image. The emotion "will be due," says Aristotle, not to the realistic imitation of the object, but "to the execution, the coloring, or some similar cause" (chap. 4). Aristotle's own analysis of images and imagery is evidence not only of their similar emotional effects and potential persuasive value, but also of the necessity for a critical awareness of affective elements common to both, such as their execution, context and structure.

Gorgias: Linking the Persuasive Power of Words and Images

Although Aristotle's advocacy of the rhetorical use of artistic appeal, and therefore of images and imagery, is grudging, for other orators of Aristotle's era, such as the Sophists, artistic proof, images, and the corresponding use of imagery were very important. Gorgias, a Sophist who lived just before Aristotle began writing, put heavy emphasis on artistic elements such as delivery, style, and artistic modes of proof. His *Encomium of Helen* illustrates how he sees the image as a potential means of persuasion. In this work, Gorgias tries to exonerate Helen of Troy of starting the Trojan War by showing how her flight to Troy with Paris could be seen as a matter of compulsion. He begins by considering the possibility that Paris raped her and argues that she could not have put up an effective resistance in such a case. He uses this argument as a premise to set up his succeeding contentions that, just as physical strength can ravish the body, so words and images can ravish one's reason. Gorgias argues that speech "has the form of necessity," and that, because it can "ravish" the mind, it is like magic or drugs in its effect on people (1972, 52). He proves this through some examples, the most central being that what Aristotle would call artistic appeal is sufficient to persuade a crowd to accept an argument for something that is logically false. Basing his reasoning on the affective power of poetry, Gorgias observes, "A single speech, written with art but not spoken with truth, bends a great crowd and persuades" (53). At this point one can recognize what Gorgias is getting at in his own speech. If a false speech relying on art—that is, on *ethos* and *pathos*—can "constrain" the soul and blind one to *logos*, if a whole crowd can be swayed against reason by "artistic" means, then, implicitly, Helen should surely be considered blameless for what her culture would consider perfidious and unreasonable behavior.

Having illustrated the persuasive power of words, Gorgias then compares this power to that of images. He points out that "frightening sights" are capable of "extinguishing and excluding thought" and thus causing madness (1972, 53–54). Hence, he reasons, we must conclude that images and words are effectively equal; they are both able to "ravish" the soul, to cause blindness to reason and law. As he says, the emotion that is created by images is "engraved upon the mind" and "is exactly analogous to what is spoken" (54). Thus, Gorgias ultimately equates the persuasive power of the image to that of words. Moreover, as one can see by his arguments and examples, the thing that makes the two equal is their effect upon the emotions. (It is notable that he does not limit susceptibility to emotions, words, and images to women such as Helen but gives examples that include all people.)

Although he spends much time discussing the specific emotion of fear, Gorgias does not limit the appeal of images to this emotion: Images, like words, can create great desire, too. He demonstrates this in the last section of his *Encomium*, in which, in reference to the physical beauty of Paris, he argues that visual images—especially beautiful works of art—can cause irresistible desire for whatever they depict: "whenever pictures perfectly create a single figure and form from many colors and figures, they delight the sight, while the creation of statues and the production of works of art furnish a pleasant sight to the eyes. Thus it is natural for the sight to grieve for some things and to long for others" (54). Gorgias uses this example concerning artworks to show how Helen of Troy's longing for the delightful sight of Paris might have forced her into fleeing with him every bit as effectively as the druglike words or physical ravishment he referred to earlier.

Besides the overt purpose of exonerating Helen, Gorgias's discussion of the power of the image is meant to make his Greek audience—who are proud of their powers of reason, who indeed consider those powers the mark of their superiority to barbarians—aware of the sway that other modes of persuasion may have over that of reason. In essence, he is doing what a modern teacher might do in helping her students dissect the emotional appeal of a visual advertisement: showing the audience the power of something they might have considered inconsequential and, in the process, making them cognizant of their own susceptibility to it. Through such a process of exposition, both Gorgias and the modern rhetorician attempt to enable their listeners to defend themselves against *and* to use this kind of power. Understanding the image, in other words, means comprehending its dichotomous possibilities: Its persuasive power might add to an argument, but its force and nonrational nature can distract one from a message's logical appeal, or its lack thereof. I will return to this idea in the later sections of this chapter.

Horace: Linking Poetry, Pictures, and Persuasion

As we have seen, Gorgias points to the emotional force of poetry when discussing the power of persuasive speech. This is not coincidental. Indeed, Richard Lanham (1991) reminds us that "rhetorical theory has . . . often in its history overlapped poetics," most clearly because of "the area where the two bodies of theory overlap—the connotative, suggestive, metaphoric use of language," but also because their purposes have so often coincided (131–132). As Lanham notes, Cicero maintained that the main functions of rhetoric were to teach, to please, and to move. Similarly, Cicero's contemporary, the Roman writer Horace, maintained the same mixture of persuasive, didactic and pleasing functions to be essential to poetry. As he argues in his *Ars Poetica*, that poet "has gained every vote who has mingled profit with pleasure by delighting the reader at once and instructing him" (1929/1978, 479).

More importantly, he is also the most famous classical source of the idea that poetry imitates visual images, though that idea may extend as far back as the fifth-century Greek poet Simonides (Adams 1971, 67). Horace's notion, expressed in his *Ars Poetica*, equates poetry and pictures (*ut pictura poesis*) (1929/1978, 481). Although this notion was not really meant to promote the power of visual imagery as much as to show how poems may be appreciated for different attributes, succeeding generations of poets and rhetoricians came to consider *ut pictura poesis* a dictum, encouraging the functional elision of words and images. The endurance of Horace's implicit idea of marrying the didactic, the poetic, and the visual provides a good example, along with Gorgias's linking of statues, desire, and persuasion, of the strength of the crossover between rhetoric, poetry and the visual arts.

The Adaptation and Blending of Horatian, Gorgian, and Aristotelian Concepts

Horace's concepts linking poetry, the visual image, and didactic purpose became especially popular among neoclassical thinkers (Adams 1971, 73). One such thinker, George Campbell (1719–1796), talks at length of the use of imagery in rhetoric, and in doing so expands both on Gorgias's implicit linking of images, rhetoric, and poetry and on Horace's linking of poetry, painting, and instructive persuasion. In Book I of his *Philosophy of Rhetoric*, Campbell asserts that "an harangue framed for affecting the hearts or influencing the resolves of an assembly, needs greatly the assistance both of intellect and of imagination," and that it is best to seize the attention of one's audience by appealing first to the imagination (1963, 2). The best way to do this, he contends, is through poetic imagery. Because Campbell considers poetry "one mode of oratory" (3), the methodology he encourages is one that connects this poetry-as-oratory directly to painting:

> The imagination is addressed by exhibiting to it a lively and beautiful representation of a suitable object. As in this exhibition, *the task of the orator* may, in some sort, be said, *like that of the painter*, to consist in imitation, the merit of the work results entirely from these two sources; dignity, as well in the subject or thing imitated, as in the manner of imitation; and resemblance, in the portrait or performance. (3, emphasis added)

As one may see from this passage, imitation of the thing depicted is paramount. There are three reasons for this.

First, Campbell believes the use of images can provide a means of comparison for an audience and thus work upon their sense of reason. Thus, there is a precedent for considering images as a form of *logical* proof: "The connexion . . . that generally subsisteth between vivacity and belief will appear less marvelous, if we reflect that there is not so great a difference between argument and illustration as is usually imagined"

(1963, 74). This is because "reasoning," as he sees it, "is but a kind of comparison" (74). A second reason that a "painter-like" exactness of imitation is important—as is a poetic ability to create these word-images—is that imagery can produce a deep and persuasive affective response, which "assumes the denomination of *pathetic*" (4). Ultimately, Campbell says, "The ideas of the poet," expressed in this painterly way, "give greater pleasure, command closer attention, operate more strongly on the passions, and are longer remembered" than ideas expressed by more mundane writers (74).

So powerful are poetic images that they may serve to provide great sway to oratory; for, "when in suitable coloring [these images are] presented to the mind, [they] do, as it were, distend the imagination with some vast conception, and quite ravish the soul" (1963, 3). Here, we see a clear debt to Gorgias, as Campbell presents imagery as irresistible to the emotions, even going so far as to use Gorgias's terminology of "ravishment." Also, Campbell's classification of imagery under two different categories of argument, logical and emotional, follows Aristotle's system of rhetoric.

Analyzing the Web-Based Image

How are Horace's and Gorgias's precedents for blending the poetic, imagistic, and oratorical and Aristotle's ideas of rhetoric applicable to modern electronic images? I would like to propose the following analytical system. Using Aristotle's notions of rhetoric as a starting point for discussing modern digitally based presentations, one can argue that images on an electronic screen can serve as a form of *logos,* or rational proof, especially when they consist of such things as charts and graphs. For as Campbell might say, such images serve as a means of comparison (of data and so forth, in modern contexts) and thus, of rational judgment. Also, in this mode, images can augment textual information via parataxis, that is, by being placed next to such information as a coordinate, supportive structure. Accordingly, and because the Web is intertextual by nature, some consideration of how well the text and graphics interrelate is important.[3] In terms of *logos,* this consideration is especially important, because digital graphics are sometimes used to replace written text.

There are also the appeals to *pathos* and *ethos* to consider. Just as Campbell, Gorgias, and Horace saw more value in the *pathetic* aspects of the image, so the persuasive value of the digital image is perhaps more evident when one considers it in terms of *pathos* and *ethos.* This is especially true because the latter of these terms has expanded, with the evolution of rhetoric, to signify the rhetor's general credibility, rather than just serving to denote moral character. As Campbell's essay exemplifies, *ethos* and *pathos* have usually been seen, in a classical sense, as dependent upon the nature of the images the speaker "draws" with words, as well as on such things as hand gestures, facial expressions, voice modulation, clothing, and other subtleties. On the Web, however, as with printed compositions, these nonverbal cues are usually absent. Thus, part of the judgment of the speaker's character and credibility becomes contingent, instead, upon the visual images she composes, chooses, and presents on the screen. Her choice of graphics and their nature, arrangement, and movement (if they are animated) not only are important to instilling the proper emotion in the audience (and thus elemental to *pathos*) but are also part of what the audience uses, consciously or unconsciously, to decide if she, and hence her presentation, are authoritative and believable (and thus integral to *ethos*).

In sum, I propose the following model, based on classical principles of rhetoric, to assess the persuasive impact of digital images:

1. Consider *logos:* How effectively do digital graphics work together with, or even replace, digital text to create an appeal to reason?
2. Consider *pathos:* As classical rhetoricians note, images are most powerful as a means of emotional appeal (which is why their cousins, metaphorical images, are so persuasive); thus, we should take into account how digital images work in concert with written text, or by themselves, to enhance the emotional appeal of digital messages. In particular, how effectively do the enhanced verisimilitude and vividness made possible by such digital innovations as 3-D, animated, computer-aided design (CAD), and interactive graphics and easily mastered, professional-looking layouts and fonts affect the emotional appeal of digital textuality? How do these enhanced graphic effects affect the reader's perception of other modes of appeal, such as *logos* and *ethos*?
3. Consider *ethos:* How effectively do digital images work in concert with written text, or by themselves, to enhance the *ethical* appeal (credibility) of the makers of digital messages? In particular, how do the enhanced verisimilitude and vividness made possible by such digital innovations as 3-D, animated, CAD, and interactive graphics and easily mastered, professional-looking layouts and fonts affect the credibility of those who author digital texts?

Test Case 1: An "Informational" Web Page

We can look at part of a Web page to see how these analytical criteria might be applied. A portion of a Web page designed to persuade people to consider getting Lasik surgery—a type of corrective surgery for the eye—is presented in figure 5.1. The series

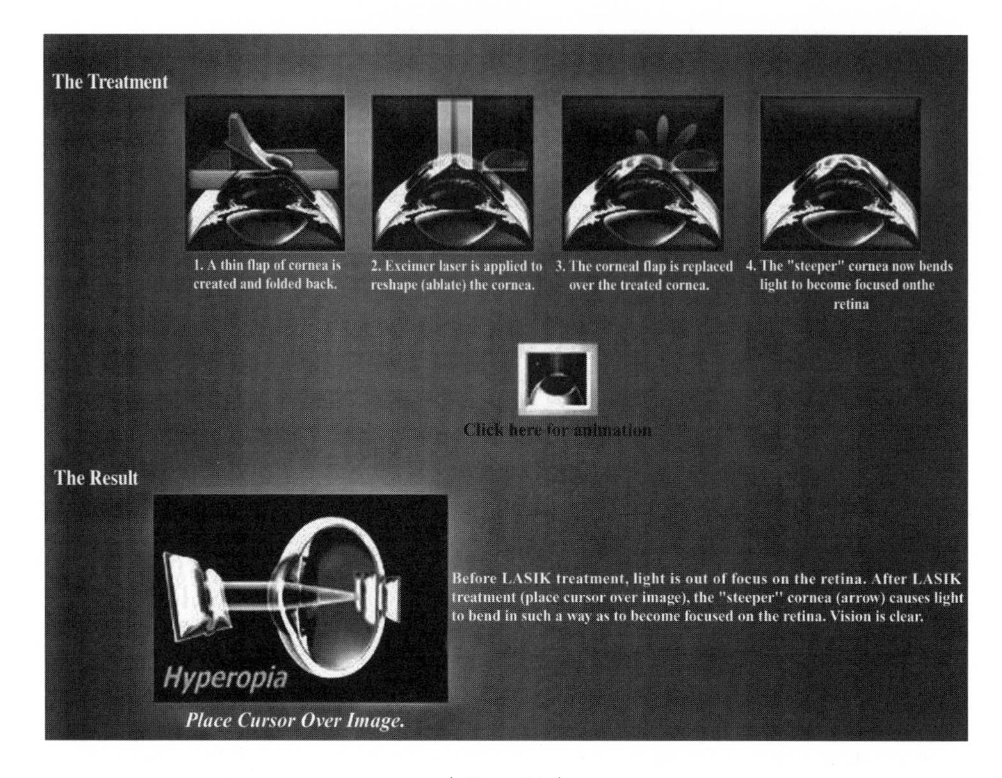

| Figure 5.1 |

Portion of a Web page designed to persuade people to consider getting Lasik surgery.

of images at the top of the figure explains the surgical procedure and is an example of an attempt to use images as *logos*. Each image is paired with a caption, but it is the image, more than the caption, that carries the clearest explanation of the procedure. Phrases like "corneal flap," "excimer laser," and "steeper cornea" are explained visually rather than verbally. The captions merely help explain what is happening in the corresponding picture. Further indication of the predominance of the image in the explanatory process is the presence of a redundant, animated version of the surgical process, using the same four images, that one can access by clicking the animation button. Nevertheless, this page presents a good sample of intertextuality and parataxis at work. Is the page persuasive? Using our criteria, I would say it is. The pictures, aided by the words, act in concert to form the *logos* of the argument: that the surgical procedure is simple, straightforward, and clean. The pictures with their captions are eye catching (which goes to *pathos*), easy to understand, and located at the top of the Web page, so that they are seen immediately upon its loading, all of which is an attempt to make the pictorially presented rationale easier for the reader to follow and is therefore important to *logos*. Additionally, the maker of this page has used both images and words to enhance the credibility (*ethos*) of the presentation. Though it is an advertisement by a doctors' office meant to generate business, it avoids any overt sales tactics: no flashy color scheme, no exhortations or radical-sounding claims. Rather, the colors are muted, with a relaxing blue as the dominant hue; the text is spare and clinical, and the illustrations have a professional, scientific appearance. This all seems calculated to instill emotions of relaxation and an *ethos* of trust in the doctors' professionalism.

Test Case 2: Of Drugs and Magic—The Problem with Ravishing Images

As we have seen, rhetorical tradition recognizes the power of images and so promotes capturing the imagination of the audience quickly by using imagistic words. The digital age allows the same purpose to be served by a return to the source of power, by a creation of "lyrical" images that delight, enrage, frighten, or excite. It seems that Web sites increasingly use this approach, and therein lies a problem, as Gorgias and his descendant Campbell warn us. The image can be seductive to the point of distraction, and this can be detrimental for both the authors and the audiences of Web sites.

About two years ago, when I began teaching basic Web design in my technical writing classes, the surprising seductiveness of some of my students' home pages got me thinking about the whole issue of using digital images as rhetorical tools. So I would like to turn to an example from one of my beginning classes to begin examining the darker side of the digital image.

Early in the semester, I always ask my students to form groups of four or five people and to devise and post a simple home page on the Internet to which they will link their succeeding assignments. This home page must be about new scientific developments related to the mind and body. Rhetorically, the aim of the home page is to convince its visitors to stay and visit the other pages that students (eventually) post regarding these new developments. I expected, when I first tried this assignment, that these sites would be relatively unsophisticated in their rhetorical appeal and, because this was an intermediate composition class, in the development of their written content—and many of them were. But some of them astonished me with their reliance on, their preoccupation with, and the attractiveness of their graphics. Figure 5.2 shows an example of one of these home pages.

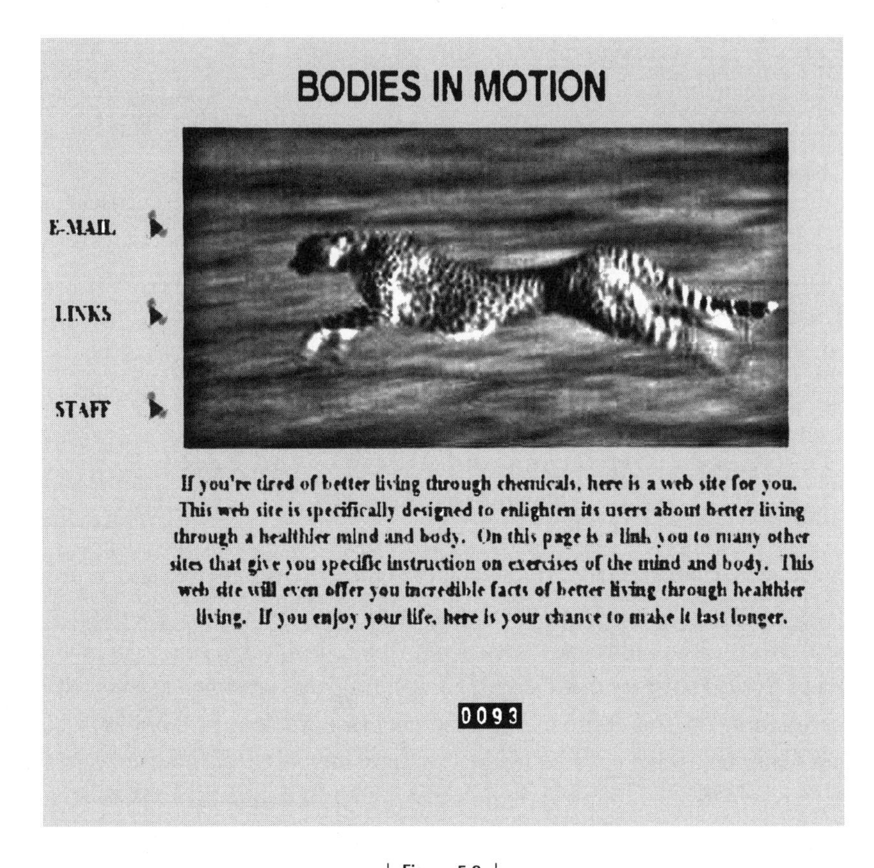

| **Figure 5.2** |

The photograph of the running cheetah on this home page effectively "advertises" the rest of the site.

The photograph of the running cheetah on this home page effectively "advertises" the rest of the site in ways that reflect the classical criteria I have been discussing. Specifically, it reflects the three classical criteria that Campbell uses: The image draws attention, invites comparison, and generates emotional response. First, the image catches the cheetah in a dramatic full sprint, which creates attentiveness. The same is true of the color scheme: Hot oranges blaze on a light background. In addition to focusing attention, the image of the cheetah could also persuade by paratactical association: The vitality of the cheetah not only invites analytical comparisons to the viewer's own vitality (or lack of it), it is also pertinent to the exercise-and-health theme of the page (if one reads and understands the text, a problem that I will return to in a moment). In terms of modern Web design, figure 5.2 includes some other elements of what might be called a good "rhetorico-graphic interface": The links are laid out well; they are placed where the eye of an English reader will focus first—in the upper left of the page. Indeed, the clean layout and flashy picture might forge an emotional impression that could linger long after the reader has left the page. The problem, however, is that the image is everything to this Web page. Though the writing shows some flair for emotional appeal in the first two sentences, the grammar, structure and clarity suffer afterward. Thus the images on this page offer no real paratactic for the words they accompany.

I had watched the students as they created this page. Fairly inexperienced at using computers, much less Web design tools, they were entranced by the image they had found and the relative ease with which they could fashion a page around it. Perhaps influenced by our immersion in a highly televisual culture, they clustered around the computer that had the image and worked on it eagerly. When I ventured over to remind them that they had to include some text, too, I watched them assign it to one reluctant person in their group, who promptly decided to put it off until later and went back to the image of the cheetah. This problem was not unique to this particular group. I have found that images—including format, layout, and even fonts for written text—take up so much of my students' attention that the idea of useful content, whether based in image or word, suffers. One mark of this is that, though ours is a technical writing class, students are often shocked when I lower their grade for bad grammar on their Web sites (this could have to do with other factors, as well, I realize, but that is another chapter in a different book). Suffice it to say that I had a difficult time getting my students to see that though a good, vivid graphical image may be enough to make a person pause on a page, it takes more than that to keep her from leaving that page; it takes a good interweaving of text and image, along with usable content.

Test Case 3: Defending Ourselves against the Dark Side of Persuasive Digital Images

This kind of experience with my students, which left me feeling somewhat like a character in *The Sorcerer's Apprentice*, has made me think that it would be a good idea to use a classically based, analytical system like the one I proposed earlier in this chapter to teach the principles of a visual rhetoric along with the elements of Web design. Teaching an awareness of the power and effects of images would provide students of Web design with a better sense not only of how to use them, but also of how to defend themselves against their power.

Indeed, there are already numerous examples of dangerous, image-centered arguments proliferating on the Web. Perhaps Aristotle said it best when he noted, as I mentioned earlier, that "the defects of the hearers" are what make images so powerful and useful to rhetoricians. The reason that my students are easily awed by fancy-looking images lies partly in the wizardry that those images convey. They look so professional, so polished, so authoritative—yet they are so easy to manipulate, and it is so easy to learn how to do so. Not surprisingly, there is a negative side to the relative ease with which one can learn how to create impressive images in digital formats. Graphics sometimes lend undue credibility to otherwise weak arguments. Even sophisticated typography and layout, graphic elements that were, before the digital age, available mainly to professional publishers, can have this result. *NBC Nightly News* ran a segment on July 19, 2000, about the fact that rumors spread on the Internet are often granted more credibility than they are due. This effect was attributed not only to the speed of rumor propagation allowed by electronic media, but also to the persuasive effect of simply seeing something in print (let alone with polished graphics) in a public venue. If an electronic document *looks* like one that has been published on paper, then, for many people, it carries the same authority. Thus, even the form and "look" of the print on a Web page can have a credibility-increasing effect. One could argue that this "print effect" occurs not only because people are conditioned to put great trust in documents that are publicly disseminated, but also because people have been socially conditioned for over 500 years to place great credibility in the typographical forms that publishers have used. Now, with electronic fonts, those forms are available to anybody.

Thus, the *ethical* effect of electronic print is at least partially a function of the fact that, like calligraphy, electronic fonts are as much art as they are signifiers of sound and words.[4] Alphabets are, in essence, abbreviations of figural metaphors (think of the evolution of Chinese ideograms, for example) and, in the digital realm, there are multitudes of fonts, designed specifically for the screen, each with its own expressive style. Web designers have already begun using these fonts and other tricks of typography for their power to affect. The Web page of a well-known hate group (figure 5.3) provides an example of how images, including typography, when interlaced cunningly with textual content, can lend undue credibility and dangerous emotional force to a site.

The methodology of this Web site is to pitch its "product" without an appeal to logic, but instead to create a sort of sublime experience, a persuasive, horrifying, visual poem. If Gorgias is right, we have a particular emotional susceptibility to fearful images. Hate mongers know this and make use of it. The rumor and the frightening image are two of their favorite devices, and this is another reason to encourage an understanding of the workings of visual rhetoric. The main rhetorical appeal here is to

| Figure 5.3 |

Portion of the Web page of a well-known hate group.

pathos, to the emotion of fear; but, ironically, it is clear from the way the images and text are constructed that the primary aim in making the Web page fearsome is to provide an aura of power, and therefore lend credibility, to the group itself. So, ultimately, the central aim of the composite image of this Web page is to affect *ethos.* The topics in the menu bar are dominated by links that provide contact and ideological information for prospective members. Most importantly, the huge image of a wolf's eyes, in conjunction with the text under it that reads, "We're everywhere you are," is meant to give a false (I hope) sense of ubiquity and power to any secret racists who visit the site. This paratactic tie between the text and image is reinforced by the interactivity between them. When one uses the mouse to roll the cursor over the image of the wolf, textual elaboration pops up regarding "Lone Wolves"—evidently a metaphor for members of the group. This pop-up text reads, "Lone Wolves are everywhere. We're in your neighborhoods, financial institutions, police departments, military, and social clubs." Not only is the message contained in the image-word combination meant to convey power and, therefore, *ethical* effect, but so is the very sophistication of the page;

the Java-based rollover function appears as a surprising, subtle—and chilling—piece of technological legerdemain.

This brings us to the secondary purpose of the page: to instill fear and intimidation in any enemies of the group who may view it; thus, its images also have a *pathetic* dimension. The typography conveys this; the font is boldface, italicized, and red to provide an image of aggressiveness, and the word "war" is written in capital letters. The Web designer's choice of a sans serif font could be seen as an attempt to accentuate the aggressive effect, as serif fonts tend to slow the eye down. The picture of a wolf's eyes and the hot, red-based colors convey aggressiveness. The kind of "published" look that this site has would have been more expensive to attain in printed pamphlets and much more difficult to disseminate before the advent of the Web. Now that it is so easy to disseminate images and messages such as the one in this Web site, it is doubly important that we pay attention to how graphics and text interact in the networked environment.

Overuse: A Limit to the Digital Image's Persuasiveness?

The persuasiveness of digital images may be limited, paradoxically, by their own power and ubiquity. Complex, graphics-heavy Web sites take a long time to load and are not very "degradable"; that is, they do not look good on older browsers and computers. This limits accessibility to these sites. But perhaps the biggest limit to the rhetorical power of graphics, even on pages where their density is not a problem, is that they can distract the reader from the logical appeal of the Web site. A recent study done at Ohio State University found that, regardless of whether a Web site's fonts and other graphic images have authoritative *form*, people had trouble understanding and focusing on the site's content. The strong *pathetic* effect of digital images can distract one from any kind of *logos* that the site might convey. One of the students in the study complained about this: "There are all these great graphics, and it takes concentration to home in and focus on the actual information" (Greenman 2000, 11). Part of this student's problem also had to do with hyperlinks, which are, technically, a type of graphical image: He found himself "struggling to digest the information on a Web page before being lured away by links to other pages."

Casual Web users are not the only ones troubled by how its hypertextual nature reinforces a focus on the *pathetic* appeal of images over any form of logical content. At least one Web designer anticipated the complaints of the students in the Ohio State study. Writing in 1998, Jeffrey Veen lamented that "designers add links by inserting harsh blue underlined scars into the patterns of the paragraphs. The result? An overbearing distraction to the reader's subconscious. Suddenly, that reader must decide: Do

I stop here and click on to this link? Do I finish the sentence and come back? Do I finish the story and scroll back to the navigation element? It's a headachy mess" (1998a, lesson 3, 1). Note that, like the classical authors we have discussed, Veen sees the attraction of the image as a factor of the "subconscious, and so of *pathos*."

The limits to digital graphics that this study exposes appear to inhere not only in the very *pathetic* power of such images, which distracted students from logical appeals contained in digital pages, but also in two other problems: the well-known difficulty of reading material on a computer screen, and an issue of image saturation. The second of these problems is particularly important: Because images in a Web environment are so emotionally appealing, they tend to be used too abundantly. As a result, as this study showed, they marginalize meaning carried in written text (which is what students in the study were asked to focus on). As Kress (1998) has pointed out, the marginalization of text by images is increasingly common in all types of printed media, and this practice has carried over to the Internet. The result appears to be an increasing reliance on digital imagery that is ever easier to manipulate and a consequent obstruction of logical appeal by emotional appeal in the digital realm. The more that this happens, the shallower the overall rhetorical appeal of digital messages becomes.[5]

Conclusion

Bizzell and Herzberg (1990) mention that Plato thought rhetoric "made a virtue of linguistic facility" by drawing attention to "the material effects of [language's] style and structure" (1165). Similarly, one could say that the integration of electronic media into the persuasive endeavor has made a virtue of digital facility by drawing attention to the material effects of graphical style and structure. When a Web site's images are especially polished, pleasing, and well arranged, its readers often cannot help but be attentive—and even impressed or moved.

The dominant effect of graphical elements may be leading to the adaptation of an advertiser-centered model of Web design, with its profusion of flashy images and persuasive appeals that work on a subconscious, emotional level, rather than on a rational one. The media critic David Shenk (1999) is skeptical about the image-rich environment of new media. He thinks that the moving image, as presented on television, and transferred thence to the Web, insidiously distracts from the substance of any message. He does not see images (he points to television as the model for the cybernetic image) as enhancing the message in any way. Shenk says, "Images captivate us effortlessly, and are difficult to filter out" (6). As an example, he uses Wim Wenders' 1991 film *Until the End of the World*, which depicts a world addicted to neurologically stimulating images.

He notes that "Wenders calls this 'the disease of images,' the problem where 'you have too many images around so that finally you don't see anything anymore'" (5). The question that arises from Shenk's discussion, with regard to teaching, is: If we encourage students to include images (especially moving ones) in their Web work, are we discouraging appeals to rational thought (*logos*)? Perhaps, if, as Bolter (1991) says, "The digitization of images inevitably . . . allows the . . . programmer, to define new contexts" for them, the real answer to this question is that we must learn to build better contexts, ones in which images work in conjunction with rational thought (72). We can begin doing this by learning to understand the rhetorical context of digital images better, and a redeployment of classical notions such as those I suggest here could prove a great help.

Notes

1. Kress (1998) notes that there is, in "information technology circles," the acute awareness of a "trend towards the visual representation of information which was formerly coded solely in language" (77); see also Brown et al 1995, as well as Tufte 1990; Lanham 1994; and Stevens 1998, especially chap.11.
2. This is a formulation that Lanham (1993) mentions in various, slightly different ways; see xii and 31, especially.
3. There are numerous sources of information on the interrelation of graphics and text on the Web itself; one that is particularly helpful because of the breadth of its articles is a Web site for designers called Webmonkey <hotwired.lycos.com/webmonkey>; see especially Frew 1997; Veen 1998a, 1998b; and Nichols 2000. My thanks, also, to Karin Kawamoto, Webmaster and technical writer, for her input on these matters.
4. See Lanham 1993, chap. 2, for a very interesting and detailed discussion of the connections between digital typography and art, as well as of art and rhetoric in the general digital realm.
5. Also, in reference to the marginalization of text by graphics, it seems that Web designers need to come up with some kind of adjustment to reduce the level of distraction hypertext links present to readers. Veen (1998a) mentions various solutions, including two interesting, low-tech ones: *re*marginalize some of the images by moving the hyperlinks to the margin of the text, so that they become like annotations, or move them all to the end of the document. He notes that these solutions have been tried by various companies, like the *New York Times*, but does not say how good the results have been.

Works Cited

Adams, Hazard. 1971. *Critical Theory since Plato.* New York: Harcourt Brace Jovanovich.

Aristotle. 1984a. *Poetics.* In *The Complete Works,* ed. Jonathan Barnes. 2 vols., 2316–2340. Princeton: Princeton University Press.

Aristotle. 1984b. *Rhetoric.* In *The Complete Works,* ed. Jonathan Barnes. 2 vols., 2152–2269. Princeton: Princeton University Press.

Bizzell, Patricia, and Bruce Herzberg, eds. 1990. *The Rhetorical Tradition: Readings from Classical Times to the Present.* Boston: St. Martin's.

Bolter, Jay David. 1991. *Writing Space: The Computer, Hypertext, and the History of Writing.* Hillsdale, NJ: Lawrence Erlbaum.

Brown, J. R., R. Earnshaw, M. Jern, and J. Vince. 1995. *Visualization: Using Computer Graphics to Explore Data and Present Information.* New York: John Wiley.

Campbell, George. 1963. *The Philosophy of Rhetoric,* ed. Lloyd F. Bitzer. Carbondale: Southern Illinois University Press.

Frew, Jim. 1997. "Design Basics." In *Webmonkey: The Web Developer's Resource.* Terra Lycos Network. Available: <http://hotwired.lycos.com/webmonkey/html/97/05/index2a.html?tw=design> (accessed February 1, 2002).

Gorgias. 1972. "Encomium of Helen," trans. George A. Kennedy. In *The Older Sophists,* ed. Rosamond Kent Sprague, 50–54. Columbia: University of South Carolina Press.

Greenman, Catherine. 2000. "Printed Page Beats PC Screen for Reading, Study Finds." *New York Times,* 10 August, p. G11.

Horace. [1929] 1978. *Ars Poetica.* In *Horace: Satires, Epistles and Ars Poetica,* trans. H. Rushton Fairclough, 442–89. Cambridge: Harvard University Press.

Kress, Gunther. 1998. "English at the Crossroads: Rethinking Curricula of Communication in the Context of the Turn to the Visual." In *Passions, Pedagogies, and 21st Century Technologies,* ed. Gail E. Hawisher and Cynthia L. Selfe, 66–88. Logan: Utah State University Press.

Lanham, Richard. 1991. *A Handlist of Rhetorical Terms,* 2nd ed. Berkeley and Los Angeles: University of California Press.

Lanham, Richard. 1993. *The Electronic Word: Democracy, Technology, and the Arts.* Chicago: University of Chicago Press.

Lanham, Richard. 1994. "The Implications of Electronic Information for the Sociology of Knowledge." *Leonardo* 27:155–163.

Nichols, Belinda. 2000. "Writing Web Documentation." In *Webmonkey: The Web Developer's Resource.* Terra Lycos Network. Available: <hotwired.lycos.com/webmonkey/00/10/index3a.html?tw=ebusiness> (accessed February 1, 2002).

Shenk, David. 1999. *The End of Patience: Cautionary Notes on the Information Revolution.* Bloomington: Indiana University Press.

Stevens, Mitchell. 1998. *The Rise of the Image, the Fall of the Word.* Oxford: Oxford University Press.

Tufte, E. R. 1990. *Envisioning Information.* Cheshire, CT: Graphics Press.

Veen, Jeffrey. 1998a. "The Foundations of Web Design." In *Webmonkey: The Web Developer's Resource.* Terra Lycos Network. Available: <http://hotwired.lycos.com/webmonkey/design/site_building/tutorials/tutorial3.html> (accessed February 1, 2002).

Veen, Jeffrey. 1998b. "Big Minds on Web Design." In *Webmonkey: The Web Developer's Resource.* Terra Lycos Network. Available: <http://hotwired.lycos.com/webmonkey/98/13/index0a.html> (accessed February 1, 2002).

Welch, Kathleen E. 1999. *Electric Rhetoric: Classical Rhetoric, Oralism, and a New Literacy.* Cambridge: MIT Press.

| 6 |
THE WORD AS IMAGE IN AN AGE OF DIGITAL REPRODUCTION

Matthew G. Kirschenbaum

"The word is an image after all," says Stuart Moulthrop (n.d.) on his Web site, and he ought to know, having produced some of the most imaginative electronic writing we've seen to date. Moulthrop offers up this axiom in the context of Web aesthetics, but it is an effective consolidation of a number of current critical convictions. For many it has become a truism that the boundaries between word and image have never been more permeable than they are now, in the midst of our "postalphabetic" era. Johanna Drucker (1995a) puts it as follows: "Contemporary life is more saturated with signs, letters, language in visual form than that of any other epoch—T-shirts, billboards, electronic marquis, neon signs, mass-produced print media—all are part of the visible landscape of daily life, especially in urban Western culture" (21). The beginnings of this broadband cultural shift to the visible spectrums of language were detectable early in the twentieth century in the typographic manifestoes of dadaism and futurism, before those experiments were conspicuously consumed by Jan Tschichold's (1995) *New Typography.* Visible language was popularized again in the rise of desktop publishing following the commercial release of the Apple Macintosh in 1984. The visible spectrum was then taken to the edge of legibility in the 1990s with the "deconstructive" graphic design of Neville Brody, David Carson, and their cohorts—before *Wired* (and countless Madison Avenue imitators) domesticated its grunge aesthetics for the new .com economy. The cumulative impact of this ongoing dialectic between avant garde and commercial consumerism should not be underestimated; for as Julian Stallabrass (1996) insists in *Gargantua*, his study of the aesthetics of mass culture, "it is obvious that the visual is the pre-eminent arena of contemporary mass culture to the extent that literacy appears to be declining in many affluent societies, not only perhaps because of declining educational resources but because the skill seems less and less relevant to many people" (4).

Not surprisingly, the last fifteen to twenty years have also witnessed a powerful persistence of vision in academic interests and agendas. The rise of new historicism and cultural studies by the late 1980s drew a vast array of visual artifacts—maps, engravings, drawings, paintings, photographs, and even video—into the fold of mainstream "textual" scholarship. The influential work of Jerome McGann and D. F. McKenzie has helped bring into focus the visual materiality of even the most commonplace texts by way of attention to their "bibliographic codes": typography, page design, illustrations, bindings, and so forth. There has been a marked rise in interest in the visual canons of major "literary" figures like William Blake and Dante Gabriel Rossetti. Meanwhile scholars such as Drucker (herself also an accomplished printer), who have long worked at the interface of the visual and the verbal arts, have seen their ideas prove increasingly influential. Contemporary poets such as Charles Bernstein, Steve McCaffery, and Susan Howe have experimented in their writing with the boundaries between word and image. Finally, the recent and rapid emergence of new media and digital culture as fields of academic inquiry has spawned a groundswell of interest in visual rhetoric and visual culture. Michael Joyce (1995), author of the landmark hypertext novel *afternoon*, foresaw this a number of years ago when he famously stated: "Hypertext is, before anything else, a visible form" (19). In short, then, a wide range of practitioners in a variety of humanities disciplines are producing an impressive array of sophisticated and compelling work, and images, as a result, have never been seen more eloquently.[1]

I myself wouldn't have it any other way, and certainly things I have written elsewhere (see, for example, Kirschenbaum 1997) align broadly with the critical positions held by figures such as Drucker and McGann. At the same time, I work regularly in one field in which the differences between texts and images have proven all but irreconcilable: applied humanities computing, where digital technologies have been used to bring a rich variety of primary source materials—from Old English manuscripts to Blake's illuminated prints, from Civil War photographs to Victorian paintings—back into popular circulation.[2] But although these electronic editions and archives have achieved considerable success in implementing sophisticated text-searching algorithms—enabling users to locate instantly a particular string among what would be the equivalent of hundreds or even thousands of printed pages—these same projects have met with considerably less success in searching and manipulating the high-quality digital facsimiles that are a large part of their attraction. The reason: As computational data structures, images differ radically and fundamentally from electronic text. Julia Flanders (1998), writing in the journal *Computers and the Humanities*, summarizes the situation this way: "The role that images currently play in electronic editions seems to move between decoration, scholarly substantiation, and bravura display. The information they contain has a more important function, intellectually, but in order for it to take on this function the image needs to occupy a different position within the electronic edition: one by which it can be processed and treated as data rather than as an encumbrance or adornment" (309).

Thus far, however, despite modest advances in fields such as computer vision and pattern recognition, images remain largely opaque to the algorithmic eyes of the machine. This discrepancy between the critical discourse about words and images in the digital age, on the one hand, and the (virtual) realities of computation, on the other, has often nagged at me. This chapter, therefore, might be described as an attempt to play devil's advocate in the face of widespread testimonials to the eloquence of the image; or put another way, it is a field report from humanities computing, where efforts to articulate images have, I think, proved illuminating.

Material evidence of the different ways words and images are represented at the computational level is readily apparent to anyone who has ever surfed the World Wide Web using a dial-up modem connection. Anyone who has had this experience knows that images always take longer to download than comparable quantities of text and may also know that the explanation lies in the fact that image files tend to be much larger than text files. The word "image," for example, requires a mere 5 bytes of memory (one byte per character) to store as American Standard Code for Information Interchange (ASCII) text.[3] The ASCII alphabet, the lingua franca of contemporary computer operating systems, consists of 128 alphanumeric characters each defined by a unique seven-bit string. (Eight bits equal one byte; a bit is the smallest unit of binary memory, either a zero or a one; so, each character in the ASCII alphabet is thus composed of its own unique sequence of 7 zeros or ones, plus 1 extraneous bit.) The same word ("image"), however, when saved as an electronic image file created using a twelve-point Courier font (figure 6.1), requires 192 bytes to represent.

In a so-called binary data format (which is how all raster-based image formats—GIF, TIFF, JPEG, and BMP, to name some of the most common—are stored), the image is saved as a bitmap, a matrix or grid in which each pixel is assigned a color value (only black or white in the case of figure 6.1, meaning that it takes just a single bit—either a one or a zero—to represent each pixel). The resulting digital file is essentially a table of these pixel values, and as we have seen, it takes 192 bytes of memory—almost

| **Figure 6.1** |

The word as image.

forty times the amount required for the ASCII file—to create such a map for this particular image. That's a trivial amount of memory by current standards, but in the real world of the Web, the numbers escalate rapidly. Color images on the Web, for example, typically range in size from 5,000 bytes (5 kilobytes) to 100,000 bytes (100 kilobytes). Anything upward of about 20 kilobytes may become tiresome for users to wait for over a modem connection that is transmitting, say, 28,800 bits per second; and if we do the math, we see that since eight bits make up one byte, a 20,000-byte (20-kilobyte) file would require approximately 5.5 seconds to transmit under optimal conditions. Recall too that many Web pages have more than just one image—some have dozens—so waits of several minutes are not uncommon for dial-up users surfing graphics-rich sites. Ironically, some of those images are likely to be bitmapped representations of text, created because the Web browser's capacity to represent visual language (banners and logos and such) in textual formats is limited; here, as Flanders (1998) notes, words function "in contradistinction to searchable *text*, as an attractive feature, something in pleasurable excess of the content" (303). The lesson in all this is that the material truths of digital reproduction exist in constant tension with the Web's siren song of the visual.

But it's not a new tension. Although the minute particulars I have been rehearsing above are the product of late-twentieth-century digital technology, the underlying issues will be familiar to any student of books and publishing. "Pictures are problems," notes Morris Eaves (2002), in the context of a discussion of the difficulties and expenses involved in printing images in university press books such as the one you are now holding (101).[4] Before exploring the status of the word as image in an age of digital reproduction, then, it may be helpful to spend a few moments looking backward.

Historically, the technologies for reproducing words and images have evolved along separate lines of development. Gutenberg's perfection of moveable type in the mid–fifteenth century involved printing from a relief surface; at the time, images were also printed in relief, from wooden blocks (woodcuts), which would be fitted into place alongside of letterpress type as a page was being set. There are also examples of woodcut books from China and Korea from around the eighth century onward, in which both words and images would be cut into the same page-sized block of wood for printing. (It's interesting to think of this process as in some ways analogous to modern page imaging in library settings—microfilm, as well as current digital imaging procedures—in which the document is reproduced as a bitmap even if it is essentially textual in nature.) The relief surface of woodcuts, however, did not lend itself to the fine detail increasingly demanded by scientific and technical illustrations. The intaglio methods (first engraving and then etching, most commonly from copper plates) that appeared from the sixteenth century onward produced far better results, particularly when coupled with tonal processes such as mezzotint and aquatint. The bifurcation between words and images is clearly evident at this point, however, with typesetting and relief printing taking place in one shop and engraved or etched images created elsewhere (in fact, a single engraving was itself often the product of multiple hands, with one artisan specializing in depicting the human figure, another in landscape effects, another in chiaroscuro, and so forth). Intaglio processes also gave rise to the need for a new type of press, the rolling press, which exerted the greater pressure required to squeeze ink out of the fine grooves of engraved and etched copper, but further partitioning the industrial means of production. Interestingly, Blake, often hailed as the first multimedia artist for his gorgeously printed illuminated poetry, developed the technique he oxymoronically called relief etching precisely so that words and images could be drawn together on the same printing surface. The invention of first lithography and then photography in the nineteenth century (and related techniques, such as lithographic transfer and the photographic halftone) altered the equation somewhat, but words and images nonetheless remained technically distinct from one another throughout every prepress process up until the rise of desktop publishing in the 1980s. Halftone screens, for example, were sometimes too fine to be printed on the coarser paper used for type, and so images would be printed separately from the text, on special paper in the center of the book.[5]

The point I want to illustrate through the above discussion is that one cannot talk about words as images and images as words without taking into account the technologies of representation upon which both forms depend. This was immediately apparent in a very practical way to persons such as type designer Zuzana Licko, who began working seriously with the Macintosh within weeks of its debut in 1984. Licko at once recognized that the limitations of the first-generation Macs (most notably their low-resolution displays, as shown in figure 6.2) were inseparable from the aesthetic components of the new fonts she was designing: "I started my venture with bitmap type

3 4 5 6 7 8 9 ? " " @

A B C D E F G H I J K L

| **Figure 6.2** |

Zuzana Licko's Oakland font.

designs, created for the coarse resolutions of the computer screen and dot matrix printer. The challenge was that because the early computers were so limited in what they could do you really had to design something special. . . . It was physically impossible to adapt 8-point Goudy Old Style to 72 dots to the inch. In the end you couldn't tell Goudy Old Style from Times New Roman or any other serif text face. . . . It is impossible to transfer typefaces between technologies without alterations because each medium has its peculiar qualities and thus requires unique designs" (Vanderlans and Licko 1993, 18, 23). Thus, what began as a material limitation in hardware and display technologies was quickly accepted, adopted, and adapted as an integral aspect of the medium's aesthetic identity, an identity that has remained iconic and intact to this day, long since the technological base has shifted beyond the crude conditions Licko describes above. Though Apple's TrueType technology, cross-licensed to Microsoft, today allows anti-aliasing, which produces smoother looking letterforms on the screen, the jaggies remain emblematic of the radical new ontologies of the medium.

It is time now to look at those radical new ontologies more closely. Numerous critics and theorists have offered up variations on Moulthrop's dictum that the word is an image after all. Mark C. Taylor and Esa Saarinen (1994), for example, claim that their new media philosophy of "imagology" insists that "the word is never simply a word but is always also an image" (Styles 3). The following account by Drucker (1995b) is more complete than most:

When a "word" is a series of keystrokes stored as information which can be mutated into another output form simply through a series of electronic moves, then the material existence of the word becomes rethought, reconfigured. Similarly, when the "word" is put into a photo-shop image as part of a pixelated tapestry so that it loses all relation to keystrokes or letters and functions just like another element in the image—subject to the same "stretch," "whirl," "twist" and other visual commands, then the status of the word as such becomes a moot point. It is merely and only—or finally and last—an image. This newly fluid boundary of identity, though prefigured by photographic media, is unique to the electronic environment.

Note that according to Drucker's formulation, the materiality of the word is *re*thought and *re*configured in electronic environments. This suggests that materiality is altered, but not eradicated. We can see some of the ways in which materiality still remains (even after Adobe Photoshop has had its way) if we consider the flip side to the "fluid boundary of identity" Drucker seems to celebrate, for although it is true that once a word is rendered as a bitmapped graphic, then image processing filters such as "stretch" and "whirl" can be brought to bear and used to their full effect, the word simultaneously ceases to be subject to automated search and retrieval technologies or to other forms of character-based manipulation (the "electronic moves" Drucker mentions in the first portion of the passage). Thus, the ASCII rendition of the word "image" can be reliably indexed and retrieved by even a very simple search engine, but the bitmapped representation of the word is, for all practical purposes, invisible to such technologies. Likewise, whereas I can easily write a shell script to replace all occurrences of the ASCII characters "i" and "m," occurring together and in that sequence, with the character "p" (but only at the start of a word, if I opt for such a constraint) there is no comparable operation at my disposal for semantically transforming the content of the word's bitmapped form.

The practical consequences of this situation are enormous and cannot be put aside merely by asserting that words are always already images. From a computational standpoint, the boundary between words and images is no more fluid than when they were transmitted with the halftone screen, the engraver's burin, or the woodcutter's knife. But humanities computing, in which scholars sensitive to the expressive power of both words and images work with their digital representations in applied settings, offers some fascinating examples of how the material differences might be leveraged. Jerome McGann, writing about his ambition to build a comprehensive hypermedia archive of the writings and pictures of Rossetti, was among the first to appreciate the full extent of the problem:

> How to incorporate digitized images into the computational field is not simply a problem that hyperediting must solve, it is a problem created by the very arrival of the possibilities of hyperediting. In my own case, the Rossetti Hypermedia Archive was begun exactly because the project forced an engagement with this problem. Those of us who were involved with the Rossetti Archive from the beginning spent virtually the entire first year working at this problem. In the end we arrived at a double approach: first, to design a structure of SGML [Standard Generalized Markup Language] markup tags for the physical features of all the types of documents contained in the Rossetti Archive (textual as well as pictorial); and second, to develop an image tool that permits one to attach anchors to specific features of digitized images. Both of these tools effectively open visual (and potentially audial) materials to the full computational power of the hyperediting environment. (2001, 69)

This "double approach" (seemingly prefigured by Rossetti's own status as a purveyor of double works, that is, poems and paintings like the "Blessed Damozel" sharing the same title and constructed in dialogic relation to each other) was to prove common throughout the 1990s: Machine-readable transcriptions of the primary-source materials, which, when tagged with descriptive encoding schemes expressed in SGML (and now Extensible Markup Language [XML]) would enable structured search and retrieval, coupled with high-quality facsimile images to record and present the physical appearance of the original document or artifact.[6] The Early American Fiction project at the University of Virginia Library's Electronic Text Center,[7] for example, is providing SGML-encoded transcriptions of over 500 works of fiction published before 1850, along with high-resolution images of *every page of every novel* acquired with a digital camera in twenty-four-bit color at 500 dots per inch. The "image tool" of which McGann speaks, now called Inote,[8] works by essentially imposing a transparent overlay on the image to which annotations can be attached (like virtual Post-it notes). Although this "opens" the image to structured search and retrieval, importantly, it does so only by outfitting the image with words—words that the machine can process and "understand," in Flanders's terms, as data.

Researchers working with primary-source humanities materials have also been drawing upon some more advanced computational techniques. Pattern matching, for example, can be applied to facsimiles of handwritten manuscripts because their pictorial content—scripted letterforms—exhibits predictable formal properties. Members of the Digital Atheneum group at the University of Kentucky, which is developing techniques for restoring and searching badly damaged Old English manuscripts,[9] describes this part of their work as follows:

> We are developing a framework for creating document-specific image processing algorithms that can locate, identify, and classify individual letterforms. In some cases a transcription may be incomplete or inaccurate because the letterforms are badly damaged or distorted and therefore difficult to identify. Although no two handwritten letters are ever exactly alike, the problem is greatly aggravated in the case of damaged or distorted text. By analyzing several representative letterforms, we hope to build computer models that can be used to perform probabilistic pattern matching of damaged letterforms. Developing such a system is prerequisite to our being able to identify fragmentary text in these manuscripts. (Seales et al. 2000, 29)

This passage is highly instructive. Computers, let us remember, are engines of formal logic. This means that they demand rule-governed behaviors to operate. In the example I used earlier, the computer could search for "i" and "m" and replace those characters with "p" (but only at the beginning of a word) because it could recognize all three characters by their bit sequences in the ASCII alphabet; because it could recognize words as discrete units (by the spaces on either side of them); and because it could identify the beginnings and the ends of words by the relative position of the characters within them. Language itself, it is strongly arguable, is a formal system; according to the *Oxford Companion to the English Language*, for example, "Language is a system in which basic units are assembled according to a complex set of rules" (McCarthur 1992, 571). The issues involved in natural-language processing notwithstanding, the computer's reliance on formal, rule-governed behaviors is broadly compatible with linguistic data. This is why historically, in the humanities, computation has focused on text-intensive applications (stylistics, corpus linguistics, and author attribution, among other areas). Such research has been underway since at least the early 1970s.

Images are another matter. Despite my earlier assertion to the contrary, one probably *could* perform a comparable search-and-replace operation on the bitmapped representation of the word "image." It would involve, however, writing a program that first identified letterforms as such by the contrast between black and white pixels and then compared individual letterforms to a library of preexisting patterns (this is essentially the way optical character recognition software works). Other problems would also have to be addressed, however: For instance, the program would have to be smart enough to first "erase" the "i" and the "m" (by flipping the pixel values to white) and then insert the "p" (by flipping the pixel values to black, but in the region of the former "m," not the former "i"—otherwise there would be a visible gap in the newly formed word). What we have, then, is clearly a much more laborious undertaking. In the ASCII alphabet, the machine has a finite character set within which formal operations can be applied. There is no comparable character set when dealing with most images, even from a semiotic standpoint. In the case of the Digital Atheneum, however, the researchers are, importantly, dealing with images of *letterforms*. Since letterforms are a class of image that exhibit unusually strong formal properties (an "A" is recognizable as an A across a very wide range of fonts and scripts), they can provide the foundation for the kind of rule-governed predictive techniques required to "repair" damaged text. The outcome is that the computer can, at least sometimes, yield otherwise unrecoverable linguistic data by exploiting the formal properties of images.

Compare this to software such as IBM's QBIC (Query By Image Content).[10] QBIC allows users to select an image (a painting, say) and then perform searches for similar images based on either color frequency ("find me paintings with the same shade of green predominating") or the spatial properties of the image ("find me other paintings with rectangular solids like this one here in the foreground"). But as John Unsworth (2000) points out, "scholars generally wonder what possessed the software designers to create

QBIC, and who in the world wants to find pictures with central blue ovals, and for what reason." The underlying issue here is what constitutes useful data in humanities settings.

At the William Blake Archive (a project in which I participate) we have developed what we believe are fairly powerful image-searching capabilities for the visual motifs in Blake's illuminated prints. Our methods rely not on pattern matching, however, but on meticulous prose descriptions of every plate's visual "components" together with an exhaustive schematic catalog of every component's visible features ("characteristics" in our parlance). Like the earlier example of the Rossetti Archive, we thus enable users to find pictures by virtue of the computer's facility for finding words. Consider the following example of the SGML[11] description for one single "component" (a shepherd) from the lower-right-hand quadrant of plate 5 of Blake's *Songs of Innocence and of Experience*, copy Z (see figure 6.3):

```
<component type="figure" location="D">

<characteristic>shepherd</characteristic>
<characteristic>male</characteristic>
<characteristic>young</characteristic>
<characteristic>short hair</characteristic>
<characteristic>crook</characteristic>
<characteristic>tights</characteristic>
<characteristic>standing</characteristic>
<characteristic>contrapposto</characteristic>
<characteristic>looking</characteristic>

<illusobjdesc>
A young, short-haired male shepherd in tights stands in contrapposto,
watching his grazing flock of sheep--perhaps looking at the sheep that
lifts its head toward him. He holds a crook in his left hand; his
purse is visible near his right knee.
</illusobjdesc>

</component>
```

Users of the archive can select characteristics in whatever combination they wish to search for from the several hundred displayed as options on our search page. A user selecting any of the characteristics included in the SGML above would thus be able to find this plate (zoomed to the lower-right-hand quadrant containing the shepherd) as a result of an image search. No doubt it is possible to find fault with the positivistic as-

| **Figure 6.3** |

William Blake's "The Shepherd" (detail).

sumptions of such a system, and our occasional critics are always right to remind us that perception is subjective (why "crook" and not "staff"?); but the more salient observation concerns the conceptual distance between a search for "male shepherds (with crooks)" and the kind of search we would expect from QBIC. Whereas it might be possible to imagine a pattern-matching algorithm that could distinguish between shepherds and sheep, how could a computer ever hope to recognize the difference between shepherds and, say, philosophers? This speaks directly to my earlier point about what kind of data are useful, and why (and how) we are looking for them in the first place. To quote the archive's editors:

As it became clear that what we wanted to use image searching to find—discriminable motifs—were already linguistically constructed, we realized we should not resist a search mechanism filtered through language. The category "sheep"—what eighteenth-century philosophic grammarians called an "abstract substantive"—is a presence generated by language. If we wish to indicate just one such creature a pointing gesture might serve as well as a word. But if we wish to think about "sheep" in general, or discuss "sheep" in all their variety throughout Blake's art, then we are already implicated in language. Further, our very perception of the world (including, of course, Blake's pictures) is in part determined by these linguistically-constructed categories. The linguistic nature of motif perception/ conception necessarily preceded any deployment of a language-based search mechanism. Whatever distortions language could bring to the perception of Blake's art had already occurred. (Eaves et al. 1999, 137)

At the Blake Archive we use the aforementioned Inote software to graft textual characteristics to specific regions of each image. This kind of image description is increasingly common in humanities computing and digital-library projects because it offers the most reliable means at our disposal for indexing visual motifs but is by no means unique to electronic environments. In fact, the history of image description, dating back to Pliny, offers numerous case studies in the fortunate fall from image to word.[12]

The examples I have been discussing have thus far all involved digital reproductions of primary-source materials: William Blake's prints, Old English manuscripts, Rossetti's poems and paintings. Humanities computing is also, however, increasingly involved in visualization: creating dynamic visual representations of large volumes of data that could not be assimilated using conventional means (such as an index or concordance). Here too the material tension between computational representations of text and image become manifest. Tony Gonzalez-Walker's Language Visualization and Research group at Cornell, for example, is developing tools to support visually certain commonplace text analysis tasks such as word frequency and proximity searches. The stated objective is as follows:

to develop a tool which allow [*sic*] a researcher to explore interactively the structures and typologies of discursive formation in large samples of textual data and develop new techniques for reading and interpreting text space. . . . The raw material for this work is ASCII text. . . . A C program converts ASCII files into IBM Data Explorer (DX) files. . . . DX then reads the text as data points on a X,Y,Z axis and plots them according to the spacing in the ASCII2DX program. (Gonzalez-Walker, n.d.)

In figure 6.4, the heavy tubular constructions are used to connect the keywords (displayed in larger and different-colored fonts), whereas the text as a whole is suspended in a series of receding planes, each virtual sheet corresponding to a page of the printed text. The resulting visual displays are striking and indeed seem to flaunt their conspicuous status as aesthetic productions as well as functional tools. As the re-

| **Figure 6.4** |

The Language Visualization and Multilayer Text Analysis project.

searchers themselves have noted with regard to visual renderings such as the one in figure 6.5: "Sometimes these images seem to say more than the standard form of analysis about the strange technology of text. As words float in space they seem to suggest a greater variety of ways in which they can interconnect and be interpreted. At times words are turned into symbols and they cut new lines of meaning and association" (Gonzalez-Walker, n.d.).

The point I wish to make, however, is that although such sentiments may seem to confirm Drucker's contention about the fluid ontological boundaries between text and image in electronic settings, it is in fact precisely the data's unequivocal computational status as *text* that permits the software to produce these remarkable visual displays. In figure 6.5, for example, the operation that suppressed the background text, leaving only certain predefined keywords on the screen (to resemble a palimpsest) is trivial by computer science standards, but only because the electronic data set is comprised of eight-bit ASCII, which, as we have seen, is eminently susceptible to just this sort of character-based manipulation. The visualizations thus demand that the underlying

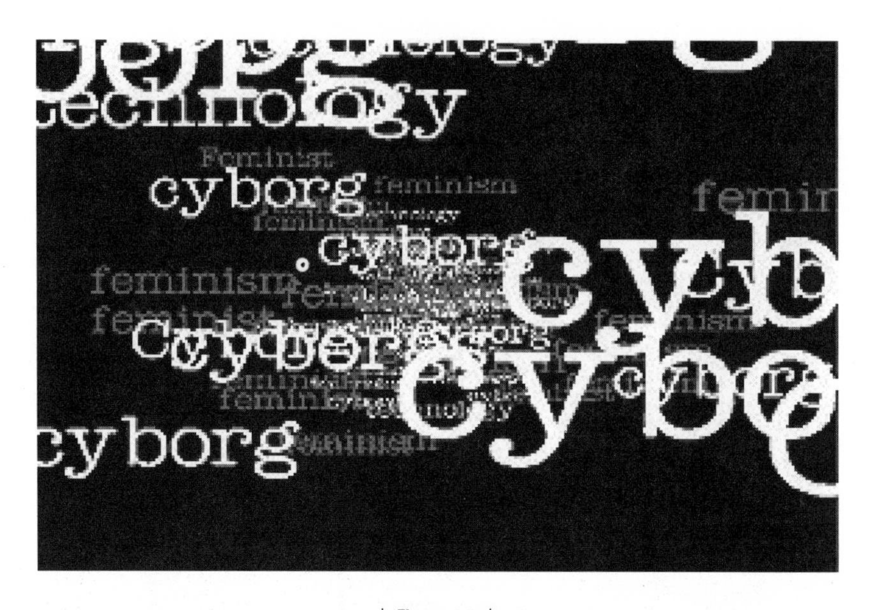

| **Figure 6.5** |

Palimpsest view from the Language Visualization and Multilayer Text Analysis project.

data be of a textual rather than a graphical nature. Here, then, the word becomes conspicuous as an image only as the result of a series of second-order operations that depend upon its primary computational ontology as machine-readable text.

My final example comes from my own work. In 1997–98 I used the Virtual Reality Modeling Language (VRML, now largely defunct) to build a three-dimensional textual installation entitled *Lucid Mapping and Codex Transformissions in the Z-Buffer.*[13] I was interested in exploring the potential of three-dimensional information landscapes (to use the late Muriel Cooper's term) as writing environments. What, for example, constitutes "narrative" in a 3-D environment? Is the distinction between linearity and nonlinearity—the crux of so much debate in hypertext theory—even relevant? Can one speak of "links" in a 3-D writing space, or does the addition of a third dimension foreground the extent to which linking is itself a flatland technology? Which of the conventions of visual and textual rhetoric translate effectively to 3-D space, and what new forms of written communication does 3-D space enable? At the time, I described the result as a t[y/o]pograhic environment, a space in which letterforms oscillated between legible signs and something like topographic contours, as in the screen shot shown in figure 6.6.

———

| **Figure 6.6** |

The t[y/o]pographies of Lucid Mapping and Codex Transformissions in the Z-Buffer.

I have observed users as they become drawn "into" this environment, visibly caught up in the process of creating and dissolving satisfying arrangements of words and letterforms through navigations of their browser. By "lucid mapping," then, I meant to suggest a sentient and directed narrative experience, assembled "on the fly" in response to changing visual and spatial conditions within a graphical environment: a mapping that then in turn alters the topography of the environment itself, and so on, thus sustaining the classic cybernetic feedback loop. Although I still think t[y/o]pography is a useful construction for accounting for this phenomenon, it's interesting in the present context to examine the workflow for the project. All of the text was created first in Adobe Photoshop, using the software's standard Type feature and some common fonts: Courier, Times New Roman, and others. Some modest use was made of various filters to lend variety to the appearance of the type, but I deliberately avoided introducing any significant distortion. The resulting image files were exported from Photoshop in GIF format, with transparent backgrounds. I then modeled a series of ultrathin three-dimensional polygons and texture-mapped them with the image files. With their transparent backgrounds, the effect was to generate two-dimensional panes of text.

Computationally, the polygons I modeled were neither images nor text: They belonged to a third class of objects known as vectors. A vector image defines, instead of a bitmap, coordinate points that are then rendered geometrically on screen. The browser draws the polygons by knowing how to interpret the mathematical definitions of their shapes and render them graphically. (Flash animations, which have largely succeeded in creating a user base on the Web where VRML failed, are all vector based.) The advantages are that the image sizes are frequently smaller than bitmaps and also that because the coordinate grid is relative and infinitely scalable, vector images do not pixelate the way raster images do when zoomed. Here is the syntax for defining a rudimentary VRML cube, texture-mapped with a checkerboard pattern, at the origin point:

```
DEF SampleCube Separator {
          Texture2 { filename "checkerboard.jpg"
                  }
                  Translation {translation 0 0 0
                  }
                  Cube {
                         width 10
                         height 10
                         depth 10
                 }
          }
```

———

Although VRML supported a basic ASCII node that allowed for the use of alphanumeric characters (labels, signposts, captions, and the like), the standard was clearly biased toward shaped objects rather than text. In *Lucid Mapping*, we have a situation in which, to achieve relatively sophisticated typographic effects, text is being computationally represented through a combination of vector constructions and graphic texture maps. Although the results are visually quite appealing, the text cannot be modified or searched, nor can it be generated or imported from preexisting electronic files. The result is that *Lucid Mapping* is precisely *not* the writing environment I set out to create. Its hold on the user is a product of the kaleidoscopic visual effects rather than any form of written interactivity.

And so we come full circle. For this is merely another manifestation of the problem that divided the reproduction of words from the reproduction of images for much of our bibliographic history. *Lucid Mapping* is not interactive because its constituent parts—its sentences and paragraphs—do not exist in any smaller combinatory units. As Eaves (1992) argues, digitization is essentially the process by which complex operations are mechanized by breaking them down into smaller operations. It is not merely a phenomenon of the current age but has been an active force throughout technological history. "As a rule of thumb, the more deeply digitization penetrates the more efficient the process becomes" (186), Eaves notes. This has immediate and tangible repercussions for our methods of reproducing texts and images. As Eaves goes on to say: "Alphabetic technology, the division of all words into a small set of uniform letters—twenty-six in the Latin alphabet, plus 10 numerals and a few "accidentals"—made efficient letterpress printing possible. Typesetters set their type in letters, not words or sentences, and a handful of little metal blocks could print every sentence" (186). *Lucid Mapping* thus behaves far more like the image on the engraver's copper plate than the words set from the compositor's cases of type. There are no reusable parts. To make something new, one must start from scratch: hammer and engrave a fresh sheet of copper, open up the software and create a fresh set of bitmaps and polygons. Seen in the context of Eaves's observations about digitization, text and images in electronic environments display a remarkable amount of resemblance to their analog counterparts. The analogy is not perfect, of course: Digital images are eminently susceptible to the rule-governed logic of computation, as Photoshop's filters demonstrate. My point is that there are significant ontological continuities with analog media that are not adequately accounted for by casual assertions about the blurred boundaries between word and image.

I have been discussing these examples from the realm of applied humanities computing not only because they illustrate the continuities between analog and digital reproduction, but also because they all furnish object lessons in the material realities of the new media. The aesthetic transformations that make digital objects so eloquent are themselves always subject to the functional constraints imposed by the material variables of computation. These include algorithms, functions, filters, file formats, data standards, memory, hardware capabilities, and user interface conventions. Understood at this level, digital artifacts are just as "real" (and tangible) as their analog counterparts. The notion that digital texts and images are infinitely fluid and malleable is an aesthetic conceit divorced from technical practice, a consensual hallucination in the same way that William Gibson's neuromantic "lines of light" delineate an imaginative ideal rather than any actual cyberspaces.

But Stuart Moulthrop (n.d.) seemed to know this all along. Here's something else he says on his Web site: "Every picture tells a story—a story of delay, complication, and increased demand on fragile systems." The context here is once again Web design, but there is a broader lesson to be learned: Images, and words, too, are almost always eloquent, not only for whatever they may signify, but also as material witnesses to their own distinctive ontologies—in any medium. I have tried to articulate those eloquences here.

Notes

1. The bibliography here is, of course, immense. For one possible mapping of the field of word and image studies, with an emphasis on the interplay between technology and representation, see <http://www.glue.umd.edu/~mgk/courses/spring2002/759/bib.html>.
2. For example, the projects indexed at <http://www.iath.virginia.edu/researchProjects.html>.
3. ASCII is recognized by the International Standards Organization (ISO 646). UNICODE, which provides a far more expansive character set, is widely regarded as the successor to ASCII.
4. The instructions to the authors of the chapters in this particular volume on the eloquence of images included the following: "Please note that we cannot plan to print color figures without greatly increasing the cost of the book." This is, of course, no reflection on the editors or the press; the situation is typical, even outside the world of scholarly publishing.
5. For an excellent introduction to the history of printing, see Twyman 1998.
6. For McGann's Rossetti Archive, see <http://www.iath.virginia.edu/rossetti>.
7. For the Early American Fiction project, see <http://etext.lib.virginia.edu/eaf>.
8. For Inote, see <http://www.iath.virginia.edu/inote>.
9. For the Digital Atheneum, see <http://www.digitalatheneum.org>.

10. For QBIC, see <http://wwwqbic.almaden.ibm.com/>. Readers may also be interested in another content-based image retrieval engine, called Blobworld: <http://elib.cs.berkeley.edu/photos/blobworld/>.
11. SGML is not a programming language. It is a descriptive metalanguage used to encode (or tag) all textual data. In the William Blake Archive, Blake's own poetry and prose, bibliographic information about a work, copy, or plate, and illustration descriptions are all prepared with SGML encoding, thus ensuring that the data are available for structured search and retrieval. XML is a somewhat simplified implementation of SGML, better suited to use on the Web.
12. For two superb studies of the tradition of image description, see Kraus 2000, 2001.
13. See <http://www.iath.virginia.edu/~mgk3k/lucid> and also Kirschenbaum 1999.

Works Cited

Drucker, Johanna. 1995a. *The Alphabetic Labyrinth: The Letters in History and the Imagination.* London: Thames and Hudson.

Drucker, Johanna. 1995b. "Synthetic Sensibilities: New Work in a Long Tradition." In *CORTEXt: A Survey of Recent Visual Poetry*, curated by Nicholas Frank and Bob Harrison, n.p. Milwaukee: Hermetic Gallery.

Eaves, Morris. 1992. *The Counter-Arts Conspiracy: Art and Industry in the Age of Blake.* Ithaca and London: Cornell University Press.

Eaves, Morris. 2002. "Graphicality: Multimedia Fables for 'Textual' Critics." In *Reimagining Textuality: Textual Studies in the Late Age of Print*, ed. Elizabeth Bergmann-Loizeaux and Neil Fraistat. Madison: University of Wisconsin Press. (99–122).

Eaves, Morris, Robert N. Essick, Joseph Viscomi, and Matthew G. Kirschenbaum. 1999. "Standards, Methods, and Objectives in the William Blake Archive." *Wordsworth Circle* 30, no. 3:135–144.

Flanders, Julia. 1998. "Trusting the Electronic Edition." *Computers and the Humanities* 31, no. 4:301–310.

Gonzalez-Walker, Tony. n.d. "Language Visualization and Multilayer Text Analysis." Cornell Theory Center. Web page. Available: <http://www.tc.cornell.edu/Visualization/contrib/cs490-95to96/tonyg/Language.Viz1.html> (accessed January 16, 2002).

Joyce, Michael. 1995. *Of Two Minds: Hypertext Pedagogy and Poetics.* Ann Arbor: University of Michigan Press.

Kirschenbaum, Matthew G. 1997. "'Through Light and the Alphabet': An Interview with Johanna Drucker. Postmodern Culture 7, no. 3 (May). Online. Available: <http://muse.jhu.edu/cgi-bin/access.cgi?uri=/journals/postmodern_culture/v007/7.3kirschenbaum/index.html>. Reprinted in Johanna Drucker, *Figuring the Word: Essays on Books, Writing, and Visual Poetics* (New York: Granary, 1998), 9–52.

Kirschenbaum, Matthew G. 1999. "Lucid Mapping: Information Landscaping and 3D Writing Spaces." *Leonard* 32, no. 4 (August):261–268.

Kraus, Kari M. 2000. "Image Description at the *William Blake Archive.* In *ALLC/ALH 2000 Proceedings, Glasgow University.* Available at <http://www.nyu.edu/its/humanities/ach_allc2001>.

Kraus, Kari M. 2001. "Images and the Language of Their Discontents." Paper presented at the ACH/ALLC 2001 Conference, New York University.

McCarthur, Tom, ed. 1992. *The Oxford Companion to English Literature.* Oxford: Oxford University Press.

McGann, Jerome. 2001. *Radiant Textuality: Literature after the World Wide Web.* New York: Palgrave.

Moulthrop, Stuart. n.d. "Web Wisdom." Web page document. Available: <http://raven.ubalt.edu/staff/moulthrop/style/pillars.html> (accessed January 16, 2002).

Seales, W. Brent, James Griffioen, Kevin Kiernan, C. J. Yuan, and Linda Cantara. 2000. "The Digital Atheneum: New Technologies for Restoring and Preserving Old Documents." *Computers in Libraries* 20, no. 2 (February):26–30.

Stallabrass, Julian. 1996. *Gargantua: Manufactured Mass Culture.* London: Verso.

Taylor, Mark C., and Esa Saarinen. 1994. *Imagologies: Media Philosophy.* London: Routledge.

Tschichold, Jan. 1995. *New Typography.* trans. Ruari McLean. Berkeley: University of California Press.

Twyman, Michael. 1998. *The British Library Guide to Printing: History and Techniques.* Toronto: University of Toronto Press.

Unsworth, John. 2000. Web page document. "The Scholar in the Digital Library." Available: <http://www.iath.virginia.edu/~jmu2m/sdl.html> (accessed January 16, 2002).

VanderLans, Rudy, and Zuzana Licko. 1993. *Emigre (the Book): Graphic Design into the Digital Realm.* New York: Van Nostrand Reinhold.

III

Perception and Knowledge in Visual and Verbal Texts

| 7 |

SAME DIFFERENCE: EVOLVING CONCLUSIONS ABOUT TEXTUALITY AND NEW MEDIA

Nancy Barta-Smith and Danette DiMarco

At some point in the second half of the twentieth century—for perhaps the first time in human history—it began to seem as if images would gain the upper hand over words.

—Mitchell Stephens, *The Rise of the Image, the Fall of the Word*

The Faces of Revolution

How is it that we speak of the visual *revolution*? Is it really, as Abigail Adams (1776/ 1973) says, that "there is a natural propensity of Humane Nature to domination"? Is revolution understood only by evaluating its break with that which came before? The epigraph that begins this chapter makes it easy to answer affirmatively. The West's propensity for dichotomies, anticipation and celebration of the recent passage of a millennial milestone, and the rapidity of technological development all encourage revolutionary talk. In this context, it is easy to see why it is tempting to make large claims about the meaning of a shift from print to visual "writing" with the advent of new media capabilities.

Of course, a preference for visual media has been evident for some time, in the popularity of mass communication vehicles such as television, video, and print advertising. Even though many have claimed that media like television cause viewers to be couch potatoes, it, and other vehicles like it, have flourished. Today, with cable and satellite dishes, viewers are able to choose among hundreds of channels. Marshall McLuhan (1965) called TV a "cold" medium because it does not extend one single sense in "high definition," as a photograph does, and provides meager information like the telephone and speech (23). But he adds that cool media are more participatory than hot media because they leave more to be "filled in or completed by the audience"

(22–23). Today, with new media, we seem to have a more embodied and literal idea of what such participation might mean.

With the fast-paced proliferation of communications technologies, it is easy to forget continuities. Our Western intellectual tradition has taught us to rely upon contraries to sharpen our analyses. For example, Geoffrey Hart (2000) writes in *Intercom* that something as basic as style guides must be reformatted for new media interactivity. The printed style guides used by writers still contain "valuable information" but they look "antiquated," and simply moving the old form on-line isn't the solution. Rather, style guides must become "dynamic," breaking with past forms. They must guide writers "through their work" as they go (12). For Hart, this means a company might construct macros or templates with embedded styles for "common tasks" (12). Most bothersome in this suggestion, aside from the standardization that English professors might frown on, is Hart's assumption that there was little interactivity between style book and writer in print and that future interactivity would take an entirely different form. Hart seems to assume that heretofore reference books have been studied and pondered like religious or philosophical tracts—perhaps memorized, rather than being constantly pulled off the shelf.

Especially in artistic contexts, in which creativity is prized over functionality, forces for change create the climate for ignoring the way that successive developments contain elements supposedly surpassed. Mitchell Stephens (1998) manifests regret that "imitation dies so hard" (50) and thinks video will truly come of age when the fast cuts of MTV displace the imitation of life in sitcoms and talk shows, etc., across other media. Laura Mulvey (1989), best known for her feminist critiques of film, says, regarding writing about photography exhibits, that such writing helps to show how a "movement can grow, change and develop, avoiding the *dangers* of fossilized repetitions and purism" (138, emphasis added). Stephens does realize the importance of gradual changes that lead up to revolutionary moments. About those who have always migrated toward the new, Stephens writes that "[i]n the end they may have as much difficulty focusing on larger, *slower* changes—such as communication revolutions—as do traditionalists" with innovation (52, emphasis added). He acknowledges that it took *centuries* for "the first two great communications revolutions . . . to produce great works" (53). Ultimately, however, Stephens believes truly innovative or revolutionary ideas will not blossom through mimicry; they will flourish when we stop imitating. Stephens's words manifest important contradictions about the way we talk about watershed moments in the development of new media.

The study of "true" imitation in comparative biology has served as the framework for our discussion below. According to Anne Russon, Robert Mitchell, Louis

Lefebvre, and Eugene Abravanel (1998), "mainstream" acceptance of the concept of "true imitation" by psychologists after Piaget caused comparative psychologists and behaviorists to abandon "Darwin's view of evolutionary continuity in favor of the dominant Western tradition of sharp, qualitative mental boundaries between humans and non-humans" (104). As a result, work in comparative biology was delayed. In contrast, evolutionary continuity would value shared characteristics across a much broader range of qualities among animals, including human ones. Niles Eldredge (2000) speaks of "punctuated equilibria" in which "the bulk of most species' histories are marked by stability" and contends that people should see this stability as a pattern or result, "not a nonresult" (22). The same may be true for our histories and discussions of new media.

To recognize continuity effectively, Eldredge would say that people must resituate their understanding of evolution from change to pattern. This pattern would be the "continuity" that Russon et al. speak of above and that has been neglected in discussions of not only human development but human invention of print and visual media as well. Not accounting much for continuity, the epigraph at the outset of this chapter (borrowed from the dust cover of Stephens's [1998] *The Rise of the Image, the Fall of the Word*) invokes a language of resistance and contrast. The title imagines textuality and visual media in a face-off in which one gets the "upper hand" over the other. The lack of attention to pattern is further stressed by the book's cover in grouping the two clauses hierarchically in terms of page design, one over the other. This chapter will evaluate the conceptual difficulties that underlie discussion of writing and new media when we suppress continuity. Applying a model set up by Russon, Mitchell, Lefebvre, and Abravanel, we argue that the claim for huge cognitive shifts from oral to print and visual literacy are overstated. Underlying all use of media, fundamental and persistent capacities for engagement and interaction, assimilation and adaptation, are shaped by embodiment and the environment. They create continuity. We attempt to initiate a conversation regarding the development of visual writing in new media that uses evolutionary concepts like hierarchical reconstruction, recapitulation, combinatorial capacity, and synchronization in evolving conclusions about the visual revolution, textuality, and new media. As Mulvey (1989) points out, "It is easier to oppose and deconstruct than to construct alternatives" (138). Nonetheless, we believe that is what needs to be done.

Identifying a True Revolution

The notion of a new media revolution almost seems ordinary. After all, new technologies like the Internet and developing media like video and film surround us. Just five years ago academic and seasoned business people alike were contemplating whether or not they would benefit from (or even have to learn) e-mail and HTML. What seemed new just a short time ago has been replaced on the part of many with an acceptance that such technologies are "here to stay" and that more developments are on the way. But how is it that we identify such technological advancements as revolutionary? How exactly is true revolution defined, and how do we really know when it is at work? In studying true revolution it seems necessary to turn to discussions of imitation, since historically imitation has been thought to be a mere lower-level component of some higher capacity for thought. Imitation has been negatively inflected. It does not hold the same status as innovation. Who wants to be labeled an imitator of a true inventor? Russon and Mitchell (1998) help identify a true revolution in that they explain historical trends in science and psychology framed in the past century. According to them, the revolutionary moment for scientists and psychologists was that which indicated a distinction between humans and nonhumans, that is, "the capacity for a symbolic versus sensorimotor reasoning, a view that can be traced to Aristotle" (104). Called "true imitation," this symbolic ability that scholars investigated was based upon observations of developmental progression and was necessarily foregrounded by more "rudimentary precursors achieved by nonhuman species" (105). Simple vision became the precursor, whereas the ability to imitate and conceptualize through memory became key to success in true imitation.

Stephens speaks of revolution in video in a way similar to that in which behaviorists and comparative psychologists have talked of true imitation, seeing a potential revolution when video shifts into a "symbolic" (Russon and Mitchell 1998, 104) or an artistic realm of metaphor as opposed to realism and storytelling. For Stephens (1998), it has "always been possible to view a fairly exact copy of a scene" (74), and this is why he sees current television and film as "superficial media" (74). It seems that such "superficial media" become comparable with sensorimotor reasoning, and that Stephens's hope for the future of video, with its flitting images and penchant for pastiche over causality and storytelling, is akin to true imitation, that which has "transcended and emerged out of rudimentary precursors" (Russon et al. 1998, 105).

Perhaps ironically, Stephens employs an Aristotelian and deliberative approach. Caught in the linearity that he so wishes to forego, he does acknowledge that he is "an inveterate reader and writer" and that his book "uses the established, wonderfully proficient medium of printed words to proclaim the potential of video, an immature and still awkward medium" (xii). Yet even if he has provided deep historical coverage and numerous claims for video's future, he is still trapped in traditionalist notions of imitation and revolution. Like Stephens, Leonard Shlain (1991), a physicist who evaluates visual artists and their breakthroughs with regard to science, argues that revolutionary visual art introduces a new worldview, something that has never been seen before (16). In effect, he agrees with Stephens that there must be a break with the past for something to be revolutionary. Shlain's opinions are further summed up in his quotation from the painter Paul Gauguin: "There are only two kinds of artists—revolutionaries and plagiarists" (quoted in Shlain 1991, 16). Celebrating imitation as combination and succession seems out of the question.

Yet identifying the actual moment when a revolutionary break occurs can be difficult. This fact has proven true in the discussion of print and visual media as well. Elizabeth Eisenstein (1998) tackled this problem when she analyzed the shift from oral, to scriptoria, to print culture. Ultimately, Eisenstein dealt with talking about these shifts (she does warn that "[t]o generalize about early printing is undoubtedly hazardous" [8]) by filling in the historical gaps to sketch a continuum of change. Of course, this strategy too can be difficult, bogging scholars down in numerous strands that, in some way, contribute to a developing, contemporary web of ideas. Eisenstein notes her own difficulties when she informs readers in a later edition of her work that her assertion that "[f]ifteenth-century book production . . . moved from scriptoria to printing shops . . . was criticized for leaving out of account a previous move from scriptoria to stationers' shops" (9). Eisenstein's point illustrates the messy nature of talking about grand shifts, and it also solidifies the claim that evolutionary characteristics override disjunctive points. Below we draw not only on Maurice Merleau-Ponty's philosophy of embodiment, but Elizabeth Tebeaux's (1997) *The Emergence of a Tradition: Technical Writing in the English Renaissance, 1475–1640* to suggest that omission of writing done in pragmatic contexts, writing such as technical writing, has obscured continuities between print and new visual media. Likewise, though Bernadette Longo (2000) expresses reservations about the development of technical writing programs in English departments in her recent history *Spurious Coin: A History of Science, Management, and Technical Writing*, the development of such programs may have influenced the absence of technical writing from discussions of the latest communication "revolution."

The Evolution of a Revolution

Much hyperbole has been uttered regarding the revolution that occurred in the changeover from the primacy of speech to the emphasis on written word. In *Orality and Literacy*, Walter Ong (1982) notes logocentrism reached its apogee in the "virtually unsurpassable example" of Ramus's dialectic or logic, in which a "one-to-one correspondence between concept, word and referent" in printed text supplanted the spoken word

(168). McLuhan (1965), his student, has a similarly dramatic take on the phonetic alphabet: "Only the phonetic alphabet makes such a sharp division in experience, giving to its user an eye for an ear, and freeing him from the tribal trance of resonating word magic and the web of kinship" (84). Oral utterance was displaced by print as the "point of departure and the model for thought" (Ong, 168), "culturally richer forms of writing" seemed left behind (McLuhan, 83), and the "magically discontinuous and traditional world of the tribal word" was sacrificed to this "cool and uniform visual medium" (83). We argue that to limit thought to word, oral or print, is itself a supplanting of all we know through the thinking apparatus of the body.[1] It is hard to believe that visual media will revolutionize how we think or act if vision has been important all along. Beginning a discussion of communication with oral language may seem natural, but it is part of a long-standing trend in philosophy and psychology to substitute mind for eye, thought for things. Merleau-Ponty (1962), reminding us of the embodiment we have so often neglected in Western philosophy, tells us that meaning already inhabits things (230).

Recent research on the brain blurs distinctions between thought and language, mind and brain, brain and body, giving us a broader vision of where meaning resides than the symbolic representation of oral or written language. Though the structure of language may form a system of distinctions, these distinctions are not arbitrary or thoroughly socially constructed. They have their meaning in relation to the sensory motor habits of the flesh. The substitution of print for the elaborate drawings of hieroglyph and ideogram could have its devastating effects only if it constituted a substitution of print for the entirety of life. Leisure that allows preoccupation with texts has been common only among a scholarly elite—initially, the mathematicians of the priestly class, Aristotle tells us in the *Metaphysics* ([1952] I.1). As scholars, we have all experienced its isolation. Perhaps it is only when one is lost in thought that the word, oral or print, looms so large that it overshadows all we know of the sensuous world through embodiment. Perhaps it is only when one is lost in thought that some words' one-to-one correspondences with a referential world can seem an inherently analytic invention of print and the shift from orality to print a shift from imitation to dissemination (Ong 1982, 168–169). It seems to us that the case for a grand shift from oral to print to visual culture may be flawed, for most oral contexts are sensory contexts, and only a few of us live among printed texts.

Merleau-Ponty believed in the primacy of our perceptual world, not orality. According to him, it is through words that the sentence comes to displace the senses and the terms "subject" and "object" become the appendages of an action verb instead of actions being "qualities" of persons and things engaged in sensorimotor experience. The "if/then" of Aristotle's propositional logic represents formally the causal forces we come to understand only through our own purposeful actions. It is only at the level of abstraction that we think of "running" apart from a person who runs. If we restore this perceptual ground as primary, contemporary visual media do not seem so innovative, at least from the point of view of a cognitive shift in the way we think or write. Moreover, logical propositions were creating one-to-one correspondences between terms and entities long before the development of either Peter Ramus's dialectic or the printing press.

Ramus's visual display, in spite of its conceptual hierarchies and reliance on writing, could be seen as calling on the spatial and affective insights of vision even though they are obviously impoverished in their sensory impact. Merleau-Ponty (1962) found visual perception filled with meaning and wrongly left behind in Piaget's theory of cognitive development, where it was associated not with cold analytic thought but the child's presymbolic life: "So much is validly meant when we talk about infantile and primitive animism: not that the child and primitive man perceive objects which they try, as Comte says, to explain by intentions or forms of consciousness—consciousness, like the object, belongs to positing thought—but because things are taken for the incarnation of what they express, and because the human significance is compressed into them and presents itself literally as what they mean" (230).[2]

The recent tendency to disparage vision explored by Martin Jay (1993) in *Downcast Eyes: The Denigration of Vision in Twentieth-Century French Thought*, in which the visual appears as inherently objectifying, owes much not only to Mulvey's analysis of the representation of women in film, but also to these assumptions about printed text's reification of language and its supplanting of the role of memory and oral techniques used to aid it. Merleau-Ponty (1964e) recognized that not all vision is an aerial view from above and afar. Vision is a vehicle for all sorts of spatial and affective relations, for distance and proximity. And if anything, Merleau-Ponty finds the latter primary, since in perception even distance draws us together through the foreground and background that we come to understand as depth through our exploration of the environment: "For [others] . . . are not fictions with which I might people my desert—offspring of my spirit and forever unactualized possibilities—but my twins or the flesh of my flesh. Certainly I do not live their life; they are definitively absent from me and I from them. But that distance becomes a strange proximity as soon as one comes back home to the perceptible world" (15).

Merleau-Ponty saw that vision provides us with a sense of our immersion in the world and the child with an animistic way of thinking that owes at least as much to the perception of dynamic form (for which we have special receptors in the visual cortex) as to the projection of intentionality, as Piaget thought. If the Ramist system of arrangement relies on the conceptual in its use of definition and partition, vision looks holistically but never sees all at once, since some of what is included in site is occluded from sight. The use of spatial layout in print can be seen as having similarities to the spatial sense we develop within the terrain that surrounds us. For instance, white space "opens up" the page, allowing us to see, just as open space in any field does. The integration of images and text we currently see in new media bears a resemblance to similar visual displays in the Renaissance. To the extent that new ways of writing provide for interactivity, they mimic our existing interactions with the visual field that surrounds us.

In Piaget, symbolic representation is seen to displace the child's reliance on visual perception in cognitive development, for the latter produces at times a childish view of the world, by which Piaget mostly meant one ignorant of causation. For Merleau-Ponty, however, we exist and navigate in a world where our body and other bodies (animate and inanimate) are themselves a kind of language shaping syntax and providing an intersubjective context prior to development of the symbolic representation that can so overtake our perceptual life. It is only subsequent to Piaget's displacement of the perceptual by the conceptual, repeating Plato, that we come to call the design features of new visual media "revolutionary" innovations.

James Mark Baldwin (1906) notes that the child first makes the distinction between person and thing because of the mobile liveliness of persons in contrast to the "dead" inanimate things that inhabit our environment (56). Doesn't the child's first grasping of an object, though far from representing understanding of the concepts of means and end needed for tool use, separate it from the rest of the context and begin to train focal vision, even before we supply a name? Moreover, Piaget notes how the child's concrete sorting operations based on visual similarities prepare him or her for their formal equivalent in language and mathematics. Even the functional "similarities" involved in choosing to categorize different objects together if used in the same context, rather than sorting by kind, represent a visual association. If visual media are new, visual cognition is not. We need not worry about our ability to write for them.

Increasingly the bodily basis of our cognitive grammar is becoming evident (Lakoff 1987; Lakoff and Johnson 1980; Johnson 1987; Heine 1997). Research in this area casts many of our theories into doubt. It might suggest, as Merleau-Ponty (1964b) does in "Eye and Mind," that it is not vision but operations that objectify in Ramist logic, since analytical thought is hierarchical, subordinating means to ends and things for thought's sake (160). We have had a tendency to mistake such objectifying manipulations for sight because purposefulness turns sight into aim, substituting ends represented in consciousness for things around us. Though print is on the page, most of us do verbalize rather than visualize the word as we read. Merleau-Ponty (1962) notes in *The Phenomenology of Perception* how the word *on the page* "promotes its own oblivion . . . from the moment I am caught up in their meaning I *lose* sight of them. The paper, the letters on it, my eyes and body are there only as the minimum setting of some invisible operation" (401, emphasis added). It is not surprising, he thinks, that under such circumstances "I reach the conclusion that because I think, I am" (400). Here it is abstraction from context, not an inherent objectification implied by trading an ear for an eye, that is the problem. Such a relationship with the page is most common among scholars.

In contrast, Merleau-Ponty (1962) says, "the speaking subject plunges into speech without imagining the words he is about to utter," terms in which he also speaks of our immersion within the environment (403; see also Merleau-Ponty 1964b, 182). Both Plato's fixed (eternal) ideas and Ramist logic (though not page design) are provinces of a causal and scientific inquiry only some adults gradually understand. This inquiry objectifies knowledge insofar as it reduces things to categories, making use of them for the sake of scientific inquiry, more than because words appear on the page. For Piaget, understanding causality and the development of a capacity for representation are intertwined, and this intertwining has something to teach us about why we so ardently wish to speak of visual revolutions.

As noted in the introduction to the chapter, we have thought humans unique among species, and from an evolutionary standpoint the development of symbolic representation has stood as the dividing line between humans and animals. Of course children only gradually acquire language; however, humans' capacity for language has been the principal evidence of symbolic thought. Piaget discovered through his conversations with children, however, that even their "common-sense" perceptions differ profoundly from those of adults. The child's thinking is based in a kind of "perceptual" rather than causal logic. For instance, a big lie is a lie about an elephant, not a lie with significant resulting repercussions. To break many dishes by accident is worse than one on purpose, since it is the perceptual "truth" that has cogency. Although such perceptions can be deemed erroneous, they can also display insights the adult seems to have forgotten. For Piaget, who saw child development through the lens of scientific modes of thought, the watershed moment of representation was not language per se but the discovery of causality, which was seen to simultaneously require and produce symbolic representation. Ironically, imitation, with all of its dreary connotations of unoriginality, became central to this particular moment. "True" imitation was "deferred" imitation—imitation in the absence of the model. Rather than a slavish conformity, imitation could then be open to creative adaptation in new environments, and its purely symbolic nature would be assured. The shift to symbolic thought represents a difference in kind as well, since, in the presence of the model, imitative action could be seen as a kind of reflex reaction resulting in merely formal similarity. If action were initiated in the absence of the model, however, it could imply purposeful use and creative adaptation: the beginning of instrumentation and scientific thought. Although imitation implies a resulting visual similitude, to imitate, psychologists thought, one has to figure out how to effect the similarity, to "copy" behavior. Affect became effect, signaling a shift to causal understanding rather than mimicry.

But such a view relies on a Cartesian split that allows for representation only in conscious thought *or* materialist determinism. Shaun Gallagher (1995) shows how once again Merleau-Ponty casts doubt on the revolutionary status of such "watershed" moments, since there is an interim form of "thought" between these alternatives: bodily schemas (227–230). With Merleau-Ponty, he agrees that the body is neither a determined reflex mechanism nor simply the handmaiden of symbolic thought. Rather, the body itself and the environment with which it interacts constrain and enable avenues of behavior by their physical structure, allowing for continuity not only between presymbolic and symbolic representation, but among species. Comparative biology is finding more continuity between species than was found when the capacity for language and representation was the criterion for a radical break. We may also find that the visual revolution is less revolutionary if we look outside the vast academic literature on the momentousness of the shift from orality to print. Similarities between nonliterary texts and new visual media provide examples of near relations in the evolution of new media.

Aping the Visual Revolution

Tebeaux (1997) notes that it is reliance on literary models that has caused us to date the beginning of technical writing as concurrent with enlightenment texts such as Francis Bacon's *Advancement of Learning*. In reality, she feels that science availed itself of a developing culture of everyday texts, which included combinations of text and visuals, used by non-elites as they became literate as the result of the proliferation of the printed word in the Renaissance. In *The Emergence of a Tradition* she demonstrates the evolution, not revolution, of technical writing, within which visuals were gradually integrated as print capability for graphics increased. She implies that this emergence of so-called scientific plain style advocated by Bacon has been overlooked because of the tendency of scholars to look to elite academic and literary contexts in their research. In doing so, she finds Ramist diagrams as the equivalent of early integration of graphics in print, not just examples of objectifying thought. Ong also based his analysis on the study of elite literary and philosophical texts and contexts.

Ironically, though Ong (1982) is correct that learning by apprenticeship did not include study in the strict sense in which we apply it to texts, his understanding of such learning as "discipleship" (9), defined as listening, repeating what is heard, mastering proverbs, etc., seems to ignore the face-to-face encounter and visual imitation in such "oral contexts." He notes that though words are "grounded in oral speech, writing tyrannically locks them into a visual field forever" (12), as if they were not already in a visual context as speech. To believe as Ong does is to be locked in discourses and to ignore all that gesture and the face to face supply in our understanding, just as context conveys the gist of unfamiliar words on the page. It is through the extreme measure of isolating the word "nevertheless" from the page and asking the reader to focus on it for sixty seconds that Ong finds contemporary readers unable to think of the word without visualizing rather than hearing it (12).

Another watershed moment in development marked by Piaget was what he called the "A not B error" (Doré and Goulet 1998, 57). He discovered that initially objects are "out of sight, out of mind." The child shows little interest in objects if they are hidden, and when she has found them in one place, she may continue to look for them there even when they have been hidden elsewhere while she is watching. Whereas Piaget saw this behavior as a failure to understand the permanence of objects and the ability to represent them when absent, recent research points to bodily schema and the persistence of visual environmental cues in understanding the child's behavior. Doré and Goulet elaborate on more recent experiments that show that the child, and some primates as well, respond to "environmental cues associated with the object at the time of its disappearance" (58). In contrast, Piaget saw such "errors" as children's solipsistic conflation of themselves with world in presymbolic thought. Doré and Goulet's studies show that the children's "memories of the past and their anticipation of the future depend entirely on the presence of relevant cues in the environment," not representation and a sense of the interiority of self-consciousness (58).[3] Doré and Goulet note that both egocentric and allocentric spatial encoding can be used to search for hidden objects before the development of symbolic representation; the organism can rely on both subjective bodily schemas and surrounding landmarks associated with objects not in memory but as present cues (66). Doré and Goulet's research recalls the same point made by Merleau-Ponty (1962) in *The Phenomenology of Perception:* that we do not leave perception behind, but that it remains a context underlying other facets of cognitive development (355), including, it seems, both oral speech and print. The command over the universe Ong and McLuhan associate with objectification in the visual field is not inherently a product of the existence of words in a visual field, for the visual field itself is not objectifying. If it were, all of our interactions with things and others would be robbed of the richness Ong and McLuhan associate with orality's participatory style.

Ironically, in a discussion similar to Tebeaux's, Ong (1982) even demonstrates what he calls the "tenaciousness of orality" in the face of print, supplying the very continuity the idea of a print *revolution* denies (115). His discussion of the displacement of orality by print and the transformation of the "all-pervasive subject" rhetoric into the three Rs (reading, writing, arithmetic) shows how looking at academic and literary genres rather than everyday contexts can bolster the tendency to favor discontinuity (finding it original) and ignore continuities. He himself notes that written rhetoric absorbed nearly all of the techniques of oral disputation. In *Understanding Media*, McLuhan (1965) associated television with radio and the telephone and all three with "the inclusive and participational spoken word," rather than a visual revolution (82). If television has gradually been assimilated into print advertising, film, video, and the Internet instead, it is important, once again, to make distinctions between academic, pragmatic, and functional contexts and entertainment. Without words, the affective appeal and message of visual images is hard to pin down. For instance, Patricia Williams (1991) notes the difficulty image-dominant ads pose for lawyers, because no claim can be clearly pointed out in such advertisements. This ambiguity is tolerable and perhaps even welcome in art and entertainment, but problematic where practical action is to be carried out effectively, as in technical documents. So it is unlikely that the chaotic cuts of MTV and television advertising will invade our daily use of new visual media in the communicative contexts of academe, business, and industry. Moreover, nearly every visual medium, perhaps with the exception of silent film, has been accompanied by oral or written speech. The contexts of work in all three areas are increasingly reliant on communication skills as technology becomes more and more synonymous with *information* technology and page design can make or break communication as readily as misspelling a word. Employers continue to emphasize the need for writing and speaking skills in our visual age.

It is clear that Tebeaux (1997), as an academic, feels a bit ill at ease about her analysis of "nonacademic" texts on such subjects as accounting, recreation, gardening, cooking, household management, medicine, and beekeeping. She asserts that there is no need to apologize for the mundane character of these works, which she describes as the "philosophical product" of the Renaissance and a form of humanistic expression insofar as they are related to the rise of individualism and the transmission of knowledge even without the privilege of higher education (3). She carefully distinguishes them from scientific texts on "causation and the nature of the universe" produced in scientific discourses (3). Nonetheless, her descriptions of these works suggest continuity between contemporary media and both print and orality. The works already reveal characteristics associated with modern technical writing—awareness of the intended audience, functionality, style to facilitate readability and utility, visual aids, and the greater precision and breadth of information available through print and graphics—thus contributing to the expansion of knowledge and the development of discourse (1). They employ a visual rhetoric at the very moment of the print revolution.

Like Ong, Tebeaux (1997) notes the way the residues of oral techniques such as rhymed stanzas persist in early written documents, suggesting that such documents may have been read to workers who needed to remember them (20). It is primarily the increased volume of information, and its technical nature, in her view, that contribute to the expansion of print and the beginning of objective presentation styles based on external standards of measurement and observed behavior and processes. She does not refer to some inherent cognitive tyranny present in text trapped on the page (30). By the early seventeenth century, roman type replaced script type to increase legibility and create spacing and wider margins, demonstrating consideration of other aspects of typography (36). Tebeaux contrasts these practical texts with their visual components to religious manuscripts adorned with aesthetic value and meant to be read "slowly and thoughtfully" (37).

In other words, Tebeaux (1997) links the *emerging* tradition of technical writing in the Renaissance with the development of what is now being called visual rhetoric in textbooks on technical writing. That visual rhetoric includes Ramist method, tabular displays, brackets, variation in font sizes, centered headings, marginal descriptors, and systems of headings to show content and organization (581). Although Tebeaux notes that Ong had suggested that writing needed and caused organization of thought as thought was translated into writing, she points out that early print books were still "clumsy, repetitious, and diffuse" in organization (40) because of their interactivity: their functional use in nonacademic contexts where orality was still extant. If Ramus's emphasis on definition and partition represents a visual execution of Aristotelian rhetoric, Tebeaux notes that it encouraged Agricolan spatialization of information and page design, which led not to the objectification of information, but to greater accessiblilty and functionality through indexes, tables of contents, white space, overviews, headings, and illustrations (53). This functionality implies a sensorimotor context not limited to print, and now heralded as *new* in new visual media such as the Internet. If Ramistic logic was replaced by late-seventeenth-century Cartesian thought, Tebeaux believes, it left its mark on technical writing through its pictures of reality and the entities Ramus called "places" (53). Merleau-Ponty would say that the use of this term demonstrates the persistence of visual perception in the lived and embodied context underlying symbolic representation. For it is in the embodied world that everything is somewhere, in contrast to the conceptual realm, whose claims reach both everywhere and nowhere.

In Tebeaux's analysis one sees continued evidence of the integration of print into a wider context than the shift from oral disputation to the three Rs in academic contexts noted by Ong (1982, 115). Technical documents varied in their inclusion of graphics and in the detail in the technical description and callouts describing visuals. The detail was greatest where silent reading would serve as the principal source of knowledge acquisition. Access to real contexts and hands-on experience made it possible to use the books as handbooks and reference aids. The more interaction with real contexts existed, the less detail was required in print. Such ideas urge caution in making grand claims about a visual revolution.

Medical books in Renaissance times varied in the extent of their anatomical drawings depending on the attitude of the Church toward dissection and resulting access or lack of access to cadavers. Even though she speaks of the emergence of page design and graphic representation during the Renaissance, Tebeaux (1997) notes that "brackets and visualized dichotomies and schematics" were used in manuscripts of the Middle Ages, making Ramus's method of establishing hierarchies of information an *added* incentive for visual display (65). Tebeaux mentions how Ramus's method was sometimes applied inappropriately to large and complex content so that the method defeated the material's purpose by forcing it to fit on a single page, reminiscent of the contemporary admonition to break Web pages up into manageable wholes (65). Although she seems to agree with Ong that in some sense such visual displays lent credence to the idea that knowledge could be "grasped, displayed, and contained within the printed page" (70), Tebeaux's metaphor seems to suggest the primacy of instrumentalism Merleau-Ponty associates with concepts, not percepts. Otherwise, flying the flag and the hieroglyph would also be cases of objectification.

Russon et al. (1998) note that it was possible to foreclose investigation of continuities with humans in comparative biology and to include only primates as near relatives through a tradition of thought in psychological literature that assumed a radical discontinuity in the capacity for representation, language, and tool use. Recent brain research, however, is blurring these distinctions and allowing us to see anew. We may find more relatives in new visual media than we have dreamed of, more than the literature on the visual revolution or print advertising, television, and film would have us believe.

———

Technical writing and communications departments have a cohistory but one fraught with conflicts over emphasis in addition to the normal blindness we acquire concerning those outside disciplinary boundaries. The emphasis of communications departments has been broadcast media, sometimes journalism, sometimes film. Disciplinary boundaries may have created a rhetoric of revolution regarding new visual media that is unwarranted if close cousins are recognized.

The Visual Evolution

In the first model of human cognitive ability, researchers have believed that "greater behavioral flexibility" arose in primates, specifically great apes, because of more developed cognitive abilities. Such a belief has been tethered to evolutionary claims as opposed to chance occurrences and can be exhibited in models of human cognitive ability called "hierarchical mental constructions" and "structural integration." The former concept, hierarchical mental constructions, involves "keeping several cognitive units in mind concurrently, combining them into new, higher-order units, then embedding new units as subunits in further cognitive compositions" (Russon et al. 1998, 128). The latter, structural integration, refers to organization. Since it is believed that cognitive success depends upon the "interplay among different types of cognitions," then there must be a term that defines the types or organizations that occur. Hierarchical mental constructions expect that various occurrences will take place. For example, it is here where "combinatorial and recursive" factors impact cognitive development. Simply put, a combinatorial mechanism sees a subject juggling many cognitive ideas simultaneously, in a sense not only combining but "recombining" them to make various "ensembles." The recursive factor entails a subject's ability to return cognitive products as "input," or to apply "cognition to existing cognitions, not just to direct sensorimotor experience" (Russon et al. 1998, 128).

During both the combinatorial and recursive experiences hierarchization does occur, so that, in a sense, information overload does not take place (Russon et al. 1998, 128). With regard to a visual evolution, real-world examples illustrate combinatorial and recursive processes. For example, the recently lauded interactive capabilities of video, CDs, and DVDs combine previously existing concepts that make them efficiently operate. If we had not for centuries already been actively reading, jumping around in books from chapter 1 to chapter 15 and to chapter 4, would we readily be as able to conceive nonlinear sequencing in current technologies represented in something as new as the World Wide Web? Similarly, without combining the knowledge of pulsating electricity, we would never have effectively known fiber optics, and without photography, we might have never known film. Likewise, the technical documents produced in business and industry undoubtedly shape the development of Web sites and Web design.

The second model of human cognitive ability, integration, sees not only hierarchization but also synchronization as crucial developing characteristics. It appears that the fallen word and the risen image are connected to the process of heterochrony, which, when linked to synchronization, was thought to be crucial in cognitive evolution. Heterochrony refers to the slowing down or speeding up of particular cognitive processes as they combine with others. Synchronization sets up the most "favorable conditions for cognitive interaction" and "offers a bridge between normally segregated cognitive structures that promotes interplay between them" (Russon et al. 1998, 130). When conditions aren't favorable this interplay may not be as fruitful as it might be; when they are favorable, there is the possibility of rapid, successive change.

For a variety of reasons, historical conditions did not enable development of the television screen into a computer screen some twenty years ago. Perhaps an explanation of this fact is that television was still in its more nascent stages or that the personal computer was not yet something readily conceived of as important by home owners, parents, and workers, since fewer schools were imagining many uses for computers beyond programming in computer science. Fewer businesses were enabling their employees to work at home. Stephens' claims that television is a static medium can be likened to statements concerning television's place in a state of neotony, the slowing of development so that "juvenile features" remain. When Stephens claims yet that the image is rising not through television but through video, he sees a potential acceleration of development in the area of video that is capable of creating new and innovative communications possibilities. New combinations that have been accelerated in sync—that have developed in tandem—might be the wiring of libraries and households so that users may retrieve articles via e-mail or on-line databases. Such wiring also lets people be transported through a hyperlink instead of going out to retrieve a book to find particular information. We think such factors as these are tools by means of which to conceive visual "evolution."

Evolving Conclusions about Textuality and New Media

Merleau-Ponty urged us to rethink Piaget's developmental model, allowing insights about perceptual life to "catch up" to the concrete and formal operations Piaget favored, contextualize them, and reconceive the adversarial relationship that operational thought inspires. For, as Merleau-Ponty (1962) says in *The Phenomenology of Perception:* "With the *cogito* begins that struggle between consciousnesses, each one of which, as

Hegel says, seeks the death of the other." But "for the struggle ever to begin, and for each consciousness to be capable of suspecting the alien presences which it negates, all must necessarily have some common ground and be mindful of their peaceful co-existence in the world of childhood" (355).

Likewise, we may find that the most revolutionary ideas about writing and new media emerge as mixtures of existing text, voice, and image, that is, as evolving combinations rather than definitive conclusions about textuality and new media. As Russon et al. (1998) found concerning imitation, in their discussion of newer research on that phenomenon that takes continuity into account, visual media are heterogeous, not homogeneous. Even purely visual media are various, not because they are dramatic departures, but because they alter existing relations and occur in different contexts. Our sociality is complex, and there are diverse clusters of technological pressures affecting both the development of new forms of print and visual combinations and their interpretation. The idea of revolution is attractive precisely because it simplifies complex interactions.

Russon et al. (1998, 104–105) discuss the premises that guided thinking about imitation in comparative biology:

- Darwin's idea of continuity was abandoned for "watershed moments" (105) between human and nonhuman that preserved the "capacity for symbolic versus sensorimotor reasoning."
- True imitation was governed by the symbolic, since comparative psychologists also focused on learning and assumed that learning theory was a proper framework for true imitation: "imitative learning, or learning to do a novel act by merely seeing it done."
- Cognitive psychology became linked to evolutionary ideas, and ontogeny was seen to recapitulate phylogeny. Imitation in animals was thus classified into levels depending on the amount of mental processing required to perform the imitation. The idea developed that imitation begins in general biological or psychological capacities that become more complex over time. True imitation creates resemblances that "transcend and emerge out of rudimentary precursors."

These premises, in turn, generated tests of true imitation that signaled the cognitive "revolution" in humans. Perhaps it is their inverse that we need now. Accept the continuity among oral, print, and visual media and search for it. Recognize the role that embodiment and presymbolic behavior continue to exercise. Create and user-test new forms of writing in real contexts. Explore the way that cross modalities can benefit the development of new forms by creating the conditions for adaptations to forms of writing ordinarily used in other places.

When we use the slang expression "same difference," we usually mean "it doesn't make a difference" or "it doesn't change what needs to be done." We have used it in our title to recognize the way that embodiment does more than just limit what we can do, for that is the way we have thought about the body in giving primacy to the mind. Rather we want to recognize what Merleau-Ponty (1968) would call our "lateral" freedom. Neither transcendent nor forever changing, the lateral takes reciprocity as its model and recognizes Piaget's mistake in considering imitation slavish. The child first imitates in the presence of the model. In such cases mental representation isn't necessary, and certainly not "figuring out how to copy," because, Merleau-Ponty thinks, the "motives" of our actions lie there surrounding us rather than behind the eyes in imagination. Meaning presents itself even without words. To this way of thinking, the best innovations in writing and new media will value existing forms, coordinating them into new arrangements, rather than celebrating their demise.

Notes

1. In "An Unpublished Text," Merleau-Ponty (1964f) notes that "[a] system of possible movements, or 'motor projects,' radiates from us to our environment. Our body is not in space like things; it inhabits or haunts space. It applies itself to space like a hand to an instrument and when we wish to move about we do not move the body as we move an object" (5). The term "motor project" comes from the work of James Mark Baldwin, whose work Piaget also knew well, as he developed his theory of motor habits and concrete motor operations.
2. Merleau-Ponty (1988) notes that even as adults we say "wine leaves the bottle," not that it is displaced from it (206). Perception gives us even inanimate objects as dynamic forms, in which every movement implies life. Persons are only special cases of this general intersubjectivity (547).
3. Merleau-Ponty (1968) noted Piaget's error regarding the child's so-called solipsism in presymbolic life. He wished to restore to credibility this life before conscious experience of interiority, which he felt Piaget named badly as "egocentrism" when in fact its focus was on the world and others, not the self (243).

Works Cited

Adams, Abigail. [1776] 1973. "Abigail Adams to Mercy Otis Warren." In *The Feminist Papers: From Adams to de Beauvoir*, ed. Alice Rossi. New York: Columbia University Press.

Andrews, Angela Giglio. 1996. "Developing Spatial Sense—A Moving Experience!" *Teaching Children Mathematics* 2:290–294.

Aristotle. 1952. *The Works of Aristotle, Volume I.* Introduction by William Benton. Chicago: Encyclopedia Britannica, 1952.

Baldwin, James Mark. 1906. *Thought and Things: A Study of the Development and Meaning of Thought or Genetic Logic.* New York: Macmillan.

Berkenkotter, Carol, and Thomas Huckin. 1995. *Genre Knowledge in Disciplinary Communication: Cognition/Culture/Power.* Hillsdale, NJ: Lawrence Erlbaum.

Doré, François, and Sonia Goulet. 1998. "The Comparative Analysis of Object Knowledge." In *Piaget, Evolution, and Development,* ed. Jonas Langer and Melanie Killen, 55–72. Hillsdale, NJ: Erlbaum.

Eisenstein, Elizabeth. 1998. *The Printing Revolution in Early Modern Europe.* Cambridge: Cambridge University Press.

Eldredge, Niles. 2000. *The Pattern of Evolution.* New York: W. H. Freeman.

Gallagher, Shaun. 1995. "Body Schema and Intentionality." In *The Body and the Self,* ed. José Bermúdez, Naomi Eilan, and Anthony Marcel, 225–244. Cambridge: MIT/Bradford Press.

Hart, Geoffrey J. S. 2000. "The Style Guide Is Dead." *Intercom* (March):12–17.

Heine, Bernd. 1997. *Cognitive Foundations of Grammar.* New York: Oxford University Press.

Jay, Martin. 1993. *Downcast Eyes: The Denigration of Vision in Twentieth Century French Thought.* Berkeley and Los Angeles: University of California Press.

Johnson, Mark. 1987. *The Body in the Mind: The Bodily Basis of Meaning, Imagination, and Reason.* Chicago: University of Chicago Press.

Lakoff, George. 1987. *Women, Fire, and Dangerous Things: What Categories Reveal about the Mind.* Chicago: University of Chicago Press.

Lakoff, George, and Mark Johnson. 1980. *Metaphors We Live By.* Chicago: University of Chicago Press.

Langer, Jonas. 1998. "Phylogenetic and Ontogenetic Origins of Cognition: Classification." In *Piaget, Evolution, and Development,* ed. Jonas Langer and Melanie Killen, 33–54. Hillsdale, NJ: Lawrence Erlbaum.

Langer, Jonas, and Melanie Killen, eds. 1998. *Piaget, Evolution, and Development.* Hillsdale, NJ: Lawrence Erlbaum Associates.

Longo, Bernadette. 2000. *Spurious Coin: A History of Science, Management, and Technical Writing.* Albany: State University of New York Press.

McLuhan, Marshall. 1965. *Understanding Media: The Extensions of Man.* New York: McGraw-Hill.

Merleau-Ponty, Maurice. 1962. *The Phenomenology of Perception,* trans. Colin Smith. London: Routledge & Kegan Paul.

Merleau-Ponty, Maurice. 1964a. "Cezanne's Doubt. In *Sense and Nonsense,* trans. with a preface by Hubert L. Dreyfus and Patricia Allen Dreyfus, 9–25. Evanston, IL: Northwestern University Press.

Merleau-Ponty, Maurice. 1964b. "Eye and Mind." In *The Primacy of Perception*, ed. James M. Edie, 159–190. Evanston, IL: Northwestern University Press.

Merleau-Ponty, Maurice. 1964c. "Everywhere and Nowhere." In *Signs*, trans. Richard C. McCleary, 126–159. Evanston, IL: Northwestern University Press.

Merleau-Ponty, Maurice. 1964d. "Indirect Language and the Voices of Silence." In *Signs*, trans. Richard C. McCleary, 39–84. Evanston, IL: Northwestern University Press.

Merleau-Ponty, Maurice. 1964e. "Introduction." In *Signs*, trans. Richard C. McCleary, 3–35. Evanston, IL: Northwestern University Press.

Merleau-Ponty, Maurice. 1964f. "An Unpublished Text." In *The Primacy of* Perception, ed. James M. Edie, 3–11. Evanston, IL: Northwestern University Press.

Merleau-Ponty, Maurice. 1968. T*he Visible and the Invisible*, ed. Claude Lefort and trans. Alphonso Lingis. Evanston, IL: Northwestern University Press.

Merleau-Ponty, Maurice. 1988. *Merleau-Ponty à la Sorbonne: Résumé de cours 1949–1952* [Summary of Courses]. Dijon-Quetigny, France: Darantière.

Mulvey, Laura. 1989. *Visual and Other Pleasures*. Bloomington: Indiana University Press.

Ong, Walter J. 1982. *Orality and Literacy: The Technologizing of the Word*. London: Routledge.

Pratarelli, Marc E., and Brenda J. Steitz. 1995. "Effects of Gender on Perception of Spatial Illusions." *Perceptual and Motor Skills* 80:625–672.

Russon, Ann E., Robert W. Mitchell, Louis Lefebvre, and Eugene Abravanel. 1998. "The Comparative Evolution of Imitation." In *Piaget, Evolution, and Development*, ed. J. Langer and M. Killen, 103–144. Mahwah, NJ: Lawrence Erlbaum.

Shlain, Leonard. 1991. *Art and Physics: Parallel Visions in Space, Time and Light*. New York: Quill.

Stephens, Mitchell. 1998. *The Rise of the Image, the Fall of the Word*. New York: Oxford University Press.

Smith, Dorothy. 1987. *The Everyday World as Problematic*. Boston: Northeastern University Press.

Smith, Dorothy. 1990. *Conceptual Practices of Power*. Boston: Northeastern University Press.

Tebeaux, Elizabeth. 1997. *The Emergence of a Tradition: Technical Writing in the English Renaissance, 1475–1640*. Amity, NY: Baywood.

Williams, Patricia. 1991. *The Alchemy of Race and Rights*. Cambridge: Harvard University Press.

8
ILLUSTRATIONS, IMAGES, AND ANTI-ILLUSTRATIONS

Jan Baetens

The specific problem I want to tackle in this chapter is the difficulty of using images as illustrations in written nonfiction texts. My approach will combine theoretical considerations and a small case study: the strategically important example of two famous collaborations of media philosopher Marshall McLuhan and graphic designer Quentin Fiore. Although I will not discuss here the more general issue of *visual culture*, I would like to mention nevertheless that I share the views developed by Nicholas Mirzoeff (1998), who both admits and refuses the notion of an absolute gap between textual ("traditional") and visual ("new") culture: Visual culture, he argues, is less to be considered a new era in the evolution of contemporary society than the result of a crisis in textual culture, whose main conceptual and ideological frameworks are no longer able to satisfactorily manage the challenges raised by the spread of new, more visual, and most of all more mass-mediatized forms of communication.[1]

The emergence of the "new," that is, the digital, media has created a situation in which the relationship of the textual and the visual is being problematized in ways even more radical than was announced by the discussions accompanying the rise of visual culture (and the supposed decline of the traditional, i.e., textual, culture). If the problem of the "supersession" of the textual, first by the visual and now by the digital, is nowadays no longer a hot issue (see Nunberg 1998), one cannot deny that the digital frame has modified the structure of the verbal, written text in at least a double manner.

First, writing seems indeed to have recovered its fundamental visuality. Reading even plain texts on a screen is not "just reading," it is experienced as "looking at characters" whose visual materiality is foregrounded to the very discomfort of the reader: How can one read if one only "sees" something, instead of immediately "seeing through" it in order to catch the meaning "behind" the visible word? Electronic writing proves to have a visual, that is, antitextual potential that should be mastered and managed and the dangers of which remind us very well of the dangers attributed to images in general (lack of clarity, *différance* of meaning, aggressive corporeality, excess of seduction, etc.).[2] Whereas in traditional graphic design, the visual is supposed to "serve" the textual, in electronic design, the text appears so badly designed that it becomes "only visual." As Patrick Lynch and Sarah Horton (1999), authors of a famous textbook on graphic design for the Web, put it: "Graphic design creates visual logic and seeks an optimal balance between visual sensation and graphic information. Without the visual impact of shape, color, and contrast, pages are graphically boring and will not motivate the viewer" (53).

It is therefore—and this is a second point—not astonishing to observe that the second aspect of the visualizing of text blocks in a digital environment, the massive introduction of illustrations, obeys a similar pattern. Confronted with the problem of the poor readability of e-writing, designers (and who has not become a designer himself or herself today?) replied by multiplying the supposedly more readable images as soon as this became possible. Owing to technological innovations, it is now easy to enrich Web pages and hypertexts with visual and graphics, first static, then animated such as *QuickTime* or *Flash* (the most recent hype at the moment when I am writing these lines). But once again, one notes that this use of visual elements does not achieve the expected result. Whereas for authors such as Jay David Bolter and Richard Grusin (1999), the combination of texts and images in Web pages is indeed an important step toward the "remediation" of earlier and still very imperfect (that is, not "realist") versions of e-writing,[3] others such as Lynch and Horton (1999) stress the chaotic use of many of these new devices that only foster visibility and prevent texts and Web sites from becoming more readable (their *bête noire* is throughout their whole book what they call "frivolous trends").

In short, the problems raised by the use of illustrations in contemporary electronic writing are not totally new. On the contrary, they continue discussions linked with the emergence of visual culture, and they reactualize the still ongoing fundamental debates on the status and position of the image itself. For all these reasons, this chapter adopts a more historical point of view, focusing on general problems of visuality and tackling a key example anterior to electronic writing as such (although everybody accepts that this example is still functioning as a role model for many designers of digital texts today). In the last section of the chapter, however, I will go back to a more direct discussion of a screen-mediated reading and writing process.

Beyond "Visual Literacy"

When we are studying the role of images used as illustrations in nonfiction texts, a topic that cannot longer be ignored by literary scholars, the very presence of these iconic elements raises immediately a preliminary question: If we live in a so-called "visual culture," shouldn't it be possible to replace completely, and once and for all, the (nonfiction) text with a visual equivalent? In fact, this question can be answered in two ways, or rather, at two levels.

First, one can think of the possibility of writing itself becoming pictogrammatic or even ideogrammatic, as can be observed in many types of everyday communication: Think, for instance, of the icons appearing in all modern computer interfaces or of the symbols taking the place of many written signals (instead of writing "exit," for example, we now use a white arrow on a green square). Yet in this regard contemporary writing, it is often said, is less a *denial* of what writing is or should be than a *rediscovery* of what it has always been but what the alphabetic culture of Western societies had, purposively or not, forgotten or ignored, namely an inherently visual way of communication:

J'ai entendu révemment Umberto Eco déclarer: "Nous avons cru que nous allions être envahis par les images, qu'elles risquaient de supplanter la parole et même de se substituer à elle; mais le succès des machines de traitement de texte nous le montre à l'évidence: nous sommes plus que jamais, au contraire, dans une civilisation de l'alphabet." Telle est bien notre fatuité. . . . Ce n'est pas l'ânonnante pauvreté de l'alphabet qui a gagné à l'apparition de l'ordinateur—elle s'en embarrasse plutôt—mais tout ce que cet alphabet a prétendu longtemps ignorer et mépriser: la polysémie graphique—celle des idéogrammes—le jeu scénique—celle des syllabaires—tant de richesses expressives dont son phonétisme n'avait que faire. Ecrire est redevenu, grâce aux petits écrans de lumière, ce qu'il était à l'origine: une lecture d'images.[4] (Christin 1999, 43–44)

Second, one can think more radically of a sequential or syntactic use of visual items, in which the very linking of strings of visual signs enables not only objects, events or ideas to be (re)produced, but also the (chrono)logical relations between them. We all know that, despite Gotthold Ephraim Lessing,[5] pictures are able tell stories, and their narrative and discursive possibilities are of course dramatically increased from the very moment one takes into account not only single images, as did Lessing and as still do most theoreticians of, for instance, photography,[6] but also sets and series of (fixed) images. The existence of very long, rich, complex and completely *mute* visual narratives, such as for instance, Marie-Françoise Plissart's (1998) photographic novel *Droit de regards* (figures 8.1 and 8.2), is then proof that the substitution of a visual discourse for a textual is perfectly possible.

The extent to which a shift from the textual to the visual can really be accomplished has been a very hot issue in many studies on visuality, most of all in the discus-

| **Figure 8.1** |

Marie-Françoise Plissart, *Droit de regards* (Right of Insight) (1985), 90.

| **Figure 8.2** |

Marie-Françoise Plissart, *Droit de regards* (Right of Insight) (1985), 91. Hawisher and Sullivan

sions on "visual literacy," that is, on our competence in making and/or reading visual elements as being part of a genuine *visual language*. Although an increasing number of authors seem to agree that the distinction between verbal and visual languages is vital (and that therefore training in visual literacy should be an *educational* and *institutional* necessity), I will here stick to the viewpoint developed by cognitive researchers more willing to put forth the idea that the opposition of verbal and visual systems is merely a superficial one, the most fundamental level being that of an "iconic," that is, material contact with the world, with its own interpretative competence preceding (and making possible) the later specification of verbal and visual systems.[7] From such a perspective, one can better understand why authors such as Paul Messaris (1994), whom I will quote here very sympathetically, do not believe at all in the existence of a specific "visual language." Discussing the thesis that visual literacy necessarily relies upon prior education, he replies as follows:

Put another way, what this argument assumes is that images, like language, are a distinct means of making sense of reality and that visual education will give students an alternative, but equally valuable, form of access to knowledge and understanding. As my students sometimes ask: "Why can't term papers and theses be done visually instead of verbally, especially in courses in visual communication?" Here, it seems to me, the equation between images and language has been taken too literally. To argue that there should be more of a balance between linguistic and pictorial education doesn't mean such a balance would enable us to do with images all the kinds of things we now do with words. Furthermore, as the line of reasoning I have been trying to develop up to this point implies, even if images do have a potential role to play as cognitive tools, it does not necessarily follow that this function is dependent on prior visual education.[8] (21–22)

This viewpoint, which I will share here, obliges us to return to the already mentioned example of mute visual narratives. When we look, for instance, at the Plissart example, we must observe that its muteness is very relative, not only because the work itself is surrounded by a mainly verbal "peritext,"[9] but also because the absence of textual complements inside the work does not free us from the necessity of explicitly verbalizing the stories we discover in it, and of course these (necessarily verbal) interpretations are active and creative transformations of the work itself (Plissart 1998). Even more telling, so to speak, would be the example of the genre invented at the very moment of the creation of *Life* in 1936. *Life*'s director, Henry Luce, launched a new way of visual communication, the "photographic essay," the aim of which was to propose a universally understandable (and thus marketable!) way of journalistic communication that

———

would no longer be dependent on text. As Colin Osman (1987) quotes him, Luce's aim was "to see and take pleasure in seeing, to see and be amazed, to see and be instructed. . . . [T]hus to see and to be shown is now the will and new expectation of half-mankind" (168). Once again, the reality is, however, rather different, for in practice the "meaning" of the images was always closely fixed and determined by editorial paratexts, as is shown with particular clarity in the case of *Life*'s own photographic essays, whose visual part is only quantitatively, not qualitatively, dominating: The meaning of the image is given by a verbal complement, not by the images themselves.

If one follows Messaris, the problem of inserting visual elements into a textual setting is no longer, provided one decides to use images anyway (I will have to return back to that point), a problem of *substitution* (how to replace words by images), but one of *correct use* (how to combine words and images in order both to avoid misunderstandings and to facilitate comprehension). Moreover, this problem is not just *technical* (although technical devices will have their say), but *tactical* and *strategic:* "how" to use images cannot be separated, indeed, from "why" they should be used.

The Reasons for a Choice

Why indeed use images in nonfiction texts? The most often quoted answers or motivations appear to have to do with the following elements, which I list here in order of increasing importance.

First of all, there is, of course, *readability*, in the material sense of the word: The presence of images helps make text blocks shorter and thus easier to grasp with the eye, while at the same time underlining the global structure of the text (the list of images functions as a kind of shortcut to the whole development of the text). Second, there is also *economy of information:* Images provide information in a way that is supposed to be more rapid, more clear, than their verbal equivalent. Third, there is the aspect of *modernity:* Images "connote" modernity, and their absence may be received as a sign of boredom, old-fashionedness, and refusal of the contemporary world and the habits of today's readership (the absence or rarity of images no longer fits the image of "seriousness" as it has long done in more high-brow publications, such as the French newspaper *Le Monde*, for instance, or academic journals such as *Critical Inquiry*, in which the number of images is steadily growing). And fourth, one should also stress, and this feature is paramount, the necessity of *networking*, studied by Bruno Latour (1987) and others. Constructivist sociologists of science such as Latour have indeed revealed that "facts" (defined as "truths" that are very largely accepted) are not given but produced and that the production of these facts is possible thanks to, among many other things, the special rhetoric devices used in scientific work and publishing. Images have their role to play in this strategy, since they reinforce the building of a network of relationships that make a proposed "fact" difficult to attack, as it takes time, money, energy, competence, and so on to be able to produce credible objections.

This situation is of course an "ideal" one. In practice, each one of these advantages may imply (and often also really implies) its negative counterparts, which it is crucial to know if one wants to exploit fully the possibilities of images. First, as far as the *readability* issue is concerned, one has, for instance, to acknowledge that in many cases the very basic rules of "envisioning information" (Tufte 1990) are cruelly ignored by many typographers, and obviously the increased layout facilities of current text-processing software and computers make this problem almost inextricable: The very fact that almost everybody has nowadays become able to layout his or her own texts, and particularly to illustrate them freely (with no other restrictions than those of copyright and, most dangerously, one's own "imagination"), engenders everywhere the sort of dramatic communication catastrophes that the use of images theoretically aims to exclude! Things are no easier when, second, *economy of information* is concerned. Indeed, if images are incorrectly or insufficiently captioned, there is always a serious risk of losing instead of transmitting information. The situation of the image is here particularly paradoxical: One the one hand, when there's no good verbal counterweight to what has been called the "hell of connotation" (Baker 1985), one will never be able to make his or her reader understand the specific meaning(s) of the image one has in mind, since it is nearly impossible to "control" what images mean; on the other hand, however, this verbal "closing" of the image reduces, of course, the proper interest of the insertion of images, since one can wonder why images are being used anyway if their role does not exceed that of the visual illustration of a caption. Third, and this element results, of course, from the difficulty of "mastering" the dizzying diversity of possible meanings of an image, there is also the risk of an "iconoclastic" reaction (for a modern rereading of iconoclasm, see Mitchell 1985): When confronted with images the meaning of which complexifies instead of simplifying the verbal reasoning, the reader can be tempted to skip or even to reject the visual dimension of the text. And finally, the networked logic of scientific texts should not be considered a guarantee of success. It always remains possible for the reader to refuse the mutual references between images and texts and to "open the black box" (to use one of Latour's metaphors). Certainly in the case of iconic, that is, mimetic, not schematic or diagrammatic, representations, the reader can decide to "check" the supposed relationships between the visual and the verbal elements of a text, and sometimes the very decision to do so makes the whole reasoning (and fact- or truth-producing logic) of the text collapse: This is what happens, for instance, when one asks questions about the reliability of the images accompanying a newspaper article (for a long time, the presence of a text and a photograph together produced a mutual reinforcement, but since our suspicion toward images has dramatically increased [Mitchell 1992], the automatic linking of verbal and visual has lost its cultural, i.e., its ideological warrant, so that the presence of images represents as much a danger as an opportunity to whoever wants to "prove" something by networking images and texts).[10]

Pictures of Texts versus Pictures as Texts

To get a better grip on these problems, more particularly on the problem of using images, photographic or not, as illustrations in nonfiction texts, I would like to propose a very simple opposition (and of course, the great number of possible uses of illustrations explains why this opposition does not aim to give a complete or definite view on the problem). In a very general way, one may assume that there exist two basic ways of illustrating. The first type of illustration tries to underline with visual means what has to be understood at the verbal level of the text (the image is then a "picture of the text," a kind of visual *synopsis*). The second type tries to exploit the level of the image as an (almost) independent feature of the whole (the image becomes then a "text" per se: it develops a logic *in itself*).[11]

The advantage of this simple schema is that it helps one to examine more easily the fundamental internal logic of the image: Once one has understood whether the image is a *picture of the text* or a *picture in itself*, it should become easier to analyze whether the specific use of the image is coherent with the use of the images supposed by the text. Furthermore, it should also help one to better understand why some types of misreadings and misunderstandings can occur. Indeed, the logic of the first type is, of course, that of the mutual reinforcement of text and image, and in this case the problems of coherence will be absolutely decisive, whereas the fundamental logic of the second type is more that of the possible fusion (oscillating between simple juxtaposition, on the one hand, and complete blurring, on the other hand) of two more or less independent structures, and in this case the problems that will be raised will mainly have to do with readability matters (in the broad sense of the word).

A more detailed analysis of a small, but strategically very important, example may be helpful here. The two famous collaborations of media philosopher Marshall McLuhan and graphic designer Quentin Fiore, *The Medium Is the Massage* (McLuhan 19697/1996a) and *War and Peace in the Global Village* (McLuhan 1968/1996b), display indeed some striking occurrences of what can happen when words and images are brought together in a logic that aims to be that of a mutual reinforcement, but in fact appears to be terribly hindered by the resistance of the image to its structural subordination to the text (or, let's say, the message of the book).

As most everyone knows, McLuhan's media theory (as "academically" defended in his famous *Understanding Media* [Mc Luhan 1964/1994], and later "fashionably" illustrated in the two pocket books made in collaboration with Quentin Fiore) had at its core a stance of technological determinism, that is, the conviction that a shift in (media) technology inevitably and automatically engenders a shift in the behavior and habits of mind of the audience. In the case of McLuhan, this conviction was mingled with the very particular belief that (media) technology would be able to dissolve its own negative effects. Although the first impact of technology destroys traditional society in a way that he calls *traumatic* (McLuhan charges the invention of the printed word with all possible evils, while strongly idealizing oral culture), the step from early technology to the new technology of television (McLuhan never witnessed today's "newer" media) was supposed to solve the very problems technology itself had created: McLuhan imagined a world to come in which *electricity*, the ultimate technology in his eyes, would undo the harmful dichotomies introduced by print's fragmentation of "life." Both McLuhan's technological determinism and his philosophical and religious "holism" have been under strong attack since the 1970s.

A decisive contribution to this discourse was Raymond Williams's (1974/1990) *Television: Technology and Cultural Form*, which closed the era of technological determinism in media studies, replacing it with the cultural studies paradigm:

Sociological and psychological studies of the effects of television, which in their limited terms have usually been serious and careful, were significantly overtaken, during the 1960s, by a fully developed theory of the technology—the medium—as determining. There had been, as we have seen, much implicit ideology in the sociological and psychological inquiries, but the new theory was explicitly ideological: not only a ratification, indeed a celebration, of the medium as such, but an attempted cancellation of all other questions about it and its uses. The work of McLuhan was a particular culmination of an aesthetic theory which became negatively, a social theory: a development and elaboration of formalism which can be seen in many fields, from literary criticism and linguistics to psychology and anthropology, but which acquired its most significant popular influence in an isolating theory of 'the media.' (126–127)

In this new cultural studies paradigm, the study of visual media is no longer determined by the mere structures and features of the "medium" but as a complex interaction among different "fields" (technology, the production system, the audience, intertextuality, etc.) in which no field is ever completely hegemonic. Television in that case is not just the medium that takes the place of older media and that forces the public as well as authors to think and behave following its own new logic, but a new technology that is shaping a new mediasphere while simultaneously being shaped by it.

And McLuhan's dream of the pacified global village, enthusiastically rediscovered by the prophets of digital world culture during the 1990s, has been rapidly denounced as a kind of innocence less charming than dangerous. Lewis Lapham's (1994) words on McLuhan's "rhetoric" can now be read as an anticipation of the antiglobalist (and thus anti-McLuhanian) turn of the reflection on world culture:

It is this mystical component of McLuhan's thought that lately has revived his reputation among the more visionary promoters of "the information Superhighway" and the Internet. Journals specific to the concerns of cyberspace (*Wired* or *The Whole Earth Review*) touch on similarly transcendent themes; the authors of the leading articles talk about the late-twentieth-century substitution of "the Icon of the net for the icon of the Atom," about the virtues of "the hive mind" (its sociability and lack of memory), about the connectedness of "all circuits, all intelligence, all things economic and ecological," about the revised definitions of the self that take account of mankind's "distributed, headless, emergent wholeness." They echo McLuhan's dicta about the redemptive powers of art and the coming to pass of a millennium in which, "where the whole man is involved, there is no work."

This heroic falls into the rhythms of what I take to be a kind of utopian blank verse, and much of it seems as overblown as the bombast arriving from Washington about the beneficence of "the New World order" and the great happiness certain to unite the industrial nations of the earth under the tent of the General Agreement on Tariffs on Trade. (xviii)

The reason the collaboration of McLuhan and Fiore is so vital is twofold. First, it gives the most readable and exciting (albeit also a very troubling and dizzying) survey of what McLuhan stood for in those years. These views may now seem outdated, despite their strong revitalization during the early 1990s, yet they still remain an essential testimony to the cultural and intellectual history of a crucial decade, the 1960s. Here, too, Lapham demonstrates that the way McLuhan develops his media theory and practice is heavily loaded with untheorized ideological stances. The blurring of boundaries between words and images is not the result of some theoretical reflection on the limits and the specific characteristics of media (as was the case in the Bauhaus avant-garde, in which one finds similar "objects"), but the "natural" consequence of the holistic state of mind created by the new electric media channeling indistinctly verbal and visual images.

The point I would like to stress here is how McLuhan and Fiore's two books play a fundamental role in the emergence in a new type of academic writing, in which word and image (or better: textuality and visuality) are merged in a kind of big McLuhanian move toward "global connectedness" and in which the principle of scholarly distance and Olympian neutrality is abandoned in favor of a more committed and more "writerly" way of writing, and thus of thinking (the term "writerly" was coined by Roland Barthes [1975] in *S/Z*, in which he opposes the traditional, linear, or *readerly* text, on the one hand, and the modern, nonlinear or *writerly* text, on the other hand). Thanks to Fiore, McLuhan's books obtain a much more visual dimension, both through the multiple use of images and through the experimental layout of the pages: *The Medium Is the Massage* can be considered the encounter of Bauhaus "New Typography" on the one hand and a new way of thinking representative of electronic culture on the other. (It is, of course, significant that the strong impulse to insert images in academic work is linked with this shift toward new ways of writing, but this link is not always imperative; on the contrary, the tendency toward more visually marked-up layout seems to be part of the essence of this new type of writing.)

The Medium Is the Massage as a Forerunner of Postmodern Academic Writing

The example of McLuhan and Fiore can be viewed as a forerunner of a new kind of postmodern academic writing in which the distinction between academic argumentation and creative writing is vanishing. *The Medium Is the Massage* has been widely followed, but also dramatically transformed, less in the material sense of the word than in the ideological and philosophical background: The innovative, "writerly," academic style has been vividly influenced by Derrida (1974), who produced some very visually experimental books himself, and has also undergone some important shifts since the emergence of hypertext theory, which, of course, has itself been associated many times with Derrideanism.[12] Although other books may in the long run appear more important for the history of media theory (Bernstein 1997 gives a good survey of this subject), the collaboration of McLuhan and Fiore, particularly in their first joint adventure, *The Medium Is the Massage*, remains important enough to use it here as the terminus a quo of a new and imaginative type of visual writing (I assume that *War and Peace in the Global Village* obeys *grosso modo* the same logic, but in a visually less interesting way).

The problem with the use of images in *The Medium Is the Massage* is in fact that of a fundamental and inescapable aporia. Indeed, the insertion of images and more generally speaking the innovative layout is at odds with the main thesis of McLuhan on modern, electronic society (which is for him not a visual but an aural society), who strangely needs images and visuality to stress the idea that this society of simultaneity, emotional involvement, and participation is radically challenging the privileges of the eye, of the ancient alphabetical society characterized by segmentation, distance, and rationality: "The dominant organ of sensory and societal orientation in pre-alphabet societies was the ear—hearing was believing. The phonetic alphabet forced the magic world of the ear to yield to the neutral world of the eye. . . . Visual space is uniform, continuous, and connected. The rational man in our Western culture is a visual man. The fact that most conscious experience has little "visuality in it is lost on him. Rationality and visuality have long been interchangeable terms, but we do not live in a primarily visual world any more" (McLuhan 1967/1996a, 44–45). On the one hand McLuhan wants to challenge the civilization of the letter, the Gutenberg alphabet, and the book and therefore is looking for a medium appropriate to this message. On the other hand, he does not manage to find this new medium (which he expected to be television) and is therefore obliged to limit his ambition to the redefinition and the modification of the ancient book form, a program he performs not by the *auralization*, but by the *visualization*, of it!

This fundamental contradiction at the heart of the McLuhan and Fiore project also explains why the images represent more a danger than a helpful tool. Indeed, since images are used to make a plea against visual thinking, they can function only in a very strange manner. The difficulty becomes very clear when one examines more closely the clash between the intended logic of the illustrations, which is the logic I called here that of the "picture of the text" (logic 1), and their performed logic, which is, at least at the level of the reader response, much closer to the logic of the "picture in itself" (logic 2), which almost automatically resists the reduction of the iconic to a merely illustrative role.

More concretely, McLuhan and Fiore are confronted in their work with a twofold difficulty. First of all, their theorizing of simultaneity and ubiquity must necessarily be in contradiction with the sequential logic of the books, whose numbered pages have to be turned one after another. Second, the large set of very heterogeneous types of images in *The Medium Is the Massage* confronts the user of the book (I will leave here aside any speculation on the way McLuhan himself has read his own book or imagined its interpretation by the reader) with an explosion of connotative meanings that can in no way be subsumed by the very clear meanings of the verbal messages of the book: Despite its clarity (but the clarity of a slogan is always treacherous, since it inevitably engenders strong ideological reactions), the accompanying text is not able to perform the two tasks of "anchorage" and "relay"[13] that are necessary for the reduction of the connotative chaos of any image to the denotative security of a message (a lot of images aiming to induce the idea of universality are for today's reader completely unacceptable for reasons of, for instance, sexism or ageism).

The solutions put forward by McLuhan and Fiore (1967/1996a) are no less ambiguous than the complete visual program of their collaboration. On the one hand, they clearly mutilate some of the most important and innovative (i.e., anti-Gutenberg) aspects of the image. For instance (and this is the way they generally handle the first of the two problems just discussed), they "sequentialize" the image: Far from inserting single images, or networks of images, they group the illustrations in oriented strings, introducing first a weak or neutral starting point and then leading toward a climax (the best known example is the sequence, pp. 26–41, on the medium as extension of the body). Such a linear use of images is undoubtedly a concession to one of the basic features of print culture, in which the diversity of simultaneous communication is broken by the temporality of logical segmentation. On the other hand, their refusal of the traditional caption system, which is the ordinary instrument for "anchorage" and "relay," gives the images a certain degree of interpretative freedom and openness, which in practice only strengthens the contradiction between the sequential ordering of a great number of the illustrations and the acceptance of a global "face-to-face" relationship between images and texts. (If the images are often "freed" from the traditional captions, they are never allowed to build up an argument by themselves: There is no trace of an attempt to employ some kind of visual language and hence visual thinking in *The Medium Is the Massage*.)

Let me repeat: The analysis of these contradictions and difficulties is not a critique of the book by McLuhan and Fiore, not only because *The Medium Is the Massage* remains a fascinating example of how verbal and visual elements can be brought together in academic writing, but also because the problems McLuhan and Fiore were facing in their work are by definition without solution, at least as long as one acknowledges that there is an opposition between textual and visual "languages." The only ways therefore, to achieve a certain "pacification" of the word and image problem are either to come back to a more traditional, and rather iconoclastic, use of images, in which the meaning of the visual and iconic elements is solidly "contained" without the limits of the accompanying text, or to redefine dramatically our conception of what a text is, that is, what a written verbal utterance is. As we all know nowadays, the first solution is very difficult to accept in a visual society, which is explicitly characterized by notions such as the "pictorial turn."[14] The second solution, however, which seems to be more feasible, is still lacking a solid theoretical background. Although it is widely accepted, at least in the academic community, that "writing" is a *space*, etc. (Bolter 1991), there is still a terrible undertheorizing of this revolutionary theory: In many cases, for instance, the analysis of writing as space, etc., is still completely indebted to alphabetical and Eurocentric models.

The Message of "Screen Thinking"

An interesting exception is that of "screen thinking," a theory developed by Anne-Marie Christin (1995) in *L'image écrite* (see also Christin 2000). For Christin, the opposition of text and image is only a superficial one, not because the verbal and the iconic both obey a same, deeper logic, but much more radically because the verbal, that is, the written word, should be considered a variant of the image. In her viewpoint, text is indeed in the first place a visual sign, and reading is *not fundamentally* different from looking.

The refusal to acknowledge such a widely accepted difference is not understandable within the framework of a classical semiotic analysis, in which one finds an almost ontological distinction between "analogical" and "nonanalogical" signs. The reason Christin postulates a convergence of the verbal and the visible has to do with her vision of writing, which starts not from the notion of sign, but from the notion of "screen," a *metaphor* she uses to introduce a Copernican-like revolution in the relationship between the sign, on the one hand, and the host medium, on the other hand. In Christin's vision, it is not the host medium—the screen, the two- or three-dimensional surface on which the signs appear—that comes first. The screen is thus not the passive receiver or vehicle of the sign, but an active context proposing a set of elements that have to be deciphered and interpreted (not simply "decoded") by a spectator. There is in Christin's theory no longer any fundamental dichotomy of screen and sign. Instead of defining the screen as the surface (2-D or 3-D), Christin makes a plea for the simultaneous emergence and mutual shaping of both elements. Without screen, no sign is even imaginable: The screen is indeed a space or a surface the framing of which is necessary for the very acknowledgment of the sign, and this sign has no existence whatsoever outside its relations with other elements within the screen and with the screen's surface. Simultaneously, no screen is imaginable without signs, no screen is ever "empty," even when there is "nothing to see," for the very appearance of a sign transforms any space or surface into a meaningful structure (which Christin calls "screen"). By positing the interaction of screen and sign, Christin escapes, of course, all possible type of technological determinism, and this is why her views on screen thinking can in no way be considered a variation on some neo-McLuhanian reflection on (television) screen culture or other theories inspired by the evolution of modern media.

According to Christin, there are many very prestigious examples of such a "screen practice," but they are not Western, that is, not alphabetic: Egyptian hieroglyphs, Chinese and Japanese ideograms, Sumerian cuneiforms—all these writings systems function as Christin postulates, but because of their "exotic" character, the Western reader and the Western theoretician have never been willing to accept either the fact that such "screen systems" are in fact the rule and not the exception, or the fact that the features of such a "screen system" appear also in our own, alphabetical system to a greater degree than is generally acknowledged. There is thus also, in Christin's writing, a global refusal of the Western, alphabetical interpretation or, better, decoding of the sign. For Christin, the basic structure of the sign is more visual than linguistic. Reading a screen is thus in the first place looking at it. Looking at a screen is not just identifying or understanding the signs put or projected on it, but visually organizing and interpreting the signs within the frame of the screen. And interpreting those screen signs is not a matter of deciphering already existing signs and sign repertoires, but a matter of subjective evaluation of emergent networks of framed visuality becoming signs on a screen. Such a way of looking becomes a way of thinking, more precisely of *screen thinking*, that is a contemporary, but dramatically enlarged, version of Arnheim's (1969) "visual thinking."

The analyses put forward by Christin, who has been working for more than twenty years in this field, always in close collaboration with historians of writing and cultural anthropologists, show, however, that it is possible, first, to redefine the writing and reading processes in a more visual sense, and second, to elucidate electronic writing and the word and image problem in computer writing in a very interesting way. Given the visual turn of our culture, the question is not, according to Christin, whether or not computerized writing systems can provide a communication system that is as strong and useful as the traditional writing systems now under attack, but whether it is a good idea to choose the alphabetical system to occupy the new writing space of the multimedia computer.

Christin's answer to this question, contrary to that of Umberto Eco (see note 4), is absolutely negative. She believes, for two reasons, that alphabetical writing systems are inappropriate to this new writing space, in which the literal and the metaphorical meaning of the screen merge (but only by coincidence, not by necessity: Screen writing is not at all the exclusive privilege of computer-screens). First, Christin argues that every alphabetical system is based upon a theory of the sign that is always, in the end, determined by the idea of a voice, in other words, by the idea of the verbal transcription of an oral utterance, and that therefore the very use of the alphabet inevitably loses the very visuality of writing enabled by digital environments. Second, she also asserts that alphabetical systems produce a split between signs and surfaces, whereas she tries to think of both aspects as a whole. The alphabet is anchored in the conviction that the sign is a unit, an independent semiotic item that can be mediated in various ways. Screen thinking, however, is rooted in the idea that the essential thing is the interaction of surface and signs.

To fit truly with the basic intuitions of screen thinking, a writing system should be based upon the visual feature of *proximity*, not upon the alphabetical principles of syntax, linearity, and coherence. Seen from the viewpoint of screen thinking, the proximity principle *dominates* the coherence, linearity, and syntax principles. Elements that are placed or seen close to one another (that is, gathered on and by the same screen) influence and determine one another, so that a text appearing near an image, for instance, may become also an image, and vice versa. The meaning is in all cases *chosen* (but not *fixed*) by the reader-spectator, and this meaning is not to be conceived in terms of *synthesis*, but in terms of *association:* The first is produced by the syntactical logic of the alphabet; the latter is achieved by the visual logic of the screen.

Obviously, Christin's screen thinking is an abstract concept, not simply a reflection on the visible and material properties of the computer screens in the new media. It considers the screen to be not so much a material surface, but a virtual tension capable of producing new meanings for readers-spectators who must adapt themselves to the complex and shifting relationships of surfaces and signs, be they materialized or not under the form of a digital computer screen. In this screen thinking, signs are not forms put or left behind on a screen, but elements disclosed on a field that the reader isolates and frames not for deciphering or decoding, but for subjective interpretation.

Given the difficulties revealed by the use of images as illustrations in verbal nonfiction texts, such a theory as Christin's can of course be very helpful, not least of all because it brings to the fore the fact that the problem or the challenge with word and image linking is not only on the side of the wildly connotative image, but also on that of the sagely denotative text.

Notes

1. "[T]he postmodern is the crisis raised by modernism and modern culture confronting the failure of its own strategy of visualizing. In other words, it is the visual crisis of culture that creates postmodernity, not its textuality. While print culture is certainly not going to disappear, the fascination with the visual and its effects that was a key feature of modernism has engendered a postmodern culture that is at its most postmodern when it is visual." (Mirzoeff 1998, 4). See Evans and Hall 1999 for a more cautious opinion.

2. This (often overtly gendered) conception of the image has become widely accepted since the publication of W. J. T. Mitchell (1985) *Iconology*. A good example of its influence can be seen in James Heffernan (1995) *Museum of Words*, on "ekphrasis."
3. This is how they articulate their overall realistic (or even hyperrealistic) credo: "[t]raditional graphic design could not account for moving images, so the internet and the worldwide web necessarily passed into a new phase when they began to deliver animation, fuller interactivity and digital video and audio" (Bolter and Grusin 1999, 200). For a critique of this cyberoptimism, see Baetens 2000 and Kirschenbaum 1999.
4. My translation: "Recently I've heard Umberto Eco make the following statement: 'We thought we would be invaded by images, that they would become more important than language and even replace it; however, the very success of the word processors brings clear evidence of the contrary: We are today more than ever in a civilization of the alphabet.' What Eco says is a symptom of our own self-satisfaction. . . . In fact, the real winner of the new digital writing devices is not our poor and silly alphabet, but everything that this alphabet has tried to ignore and to despise for such a long time: graphic polysemy—as for instance in the ideograms— performative play—as for instance in the manuals for young readers—in short all the expressive richness the phonetic model of our writing did not care about. Writing has become once again, thanks to the small illuminated screens, what it was originally: the reading of images."
5. In his *Laocoon* (1766), Gotthold Ephraim Lessing develops a still very influential *anti–ut pictura poesis* theory that heavily insists on the "semiotic" specificity of each medium. For a series of contemporary rereadings, see Burwick 1999.
6. Photography theory is indeed the segment of visual studies in which the resistance of sequentiality remains strongest. For other voices, see Ribière 1995.
7. See Meunier 1998. Meunier's survey of the question is directly inspired by cognitive research done by Langacker (1987) and Lakoff (1984).
8. Of course, Messaris's skepticism toward a maximalist view of visual literacy doesn't mean at all that he refuses visual education; quite the contrary. But what his book achieves is a redefinition of visual education's goals, which become simultaneously more narrow and much more precise: Visual education in his view should tend to (a) heighten the viewer's capacity for *aesthetic* appreciation, and (b) increase the viewer's awareness of visual *manipulation*. Messaris's (1997) second book, *Visual Persuasion*, continues this thinking, but not without a very healthy self-criticism (see, for instance, p. 218 on the limited results of present training in visual literacy).
9. By "peritext" (or more globally "paratext"), Gérard Genette (1987) understands all verbal and/or visual elements that "surround" the text at its margins, without really being part of it (for instance: title, name of the author, illustrations, etc.).
10. For a good example of the way "indexical" photographs have to be translated in "iconic" drawings and vice versa, see Lynch 1991.

---

11. One can wonder already why there is no room for the third option which comes immediately in mind: the text as an illustration of the images. Since I do not retain here the possibility that visual images can be languages, I would like to argue here that this possibility is more a variant of the first type, than an autonomous third type.
12. I will not discuss here this point, which would require an examination in greater detail than space permits. George Landow's (1997) successful marriage (at least institutionally speaking) of critical theory and Derrideanism, *Hypertext 2.0*, is probably an example as problematic as the one by McLuhan and Fiore that I am examining here.
13. In the sense of these words coined by Roland Barthes in his early writings on photography (collected with others in *Image-Music-Text* [1977]).
14. For a global introduction to this concept, see W. J. T. Mitchell 1994: "Whatever the pictorial turn is, then, it should be clear that it is not a return to naive mimesis, copy or correspondence theories of representation, or a renewed metaphysics of pictorial 'presence': it is rather a postlinguistic, postsemiotic rediscovery of the picture as a complex interplay between visuality, apparatus, institutions, discourse, bodies, and figurality. It is the realization that spectatorship (the look, the gaze, the glance, the practices of observation, surveillance, and visual pleasure) may be as deep a problem as various forms of reading (decipherment, decoding, interpretation, etc.) and that visual experience or 'visual literacy' might not be fully explicable on the model of textuality" (16).

Works Cited

Arnheim, Rudolf. 1969. *Visual Thinking*. Berkeley: University of California Press.

Baetens, Jan. 2000. "A Critique of Cyberhybrid-Hype." In *The Future of Cultural Studies*, ed. Jan Baetens and José Lambert, 153–171. Leuven, Belgium: Leuven University Press.

Baker, Steve. 1985. "The Hell of Connotation." *Word & Image* 1, no. 2:164–175.

Barthes, Roland. 1977. *Image-Music-Text*, trans. Stephen Heath. London: Fontana/Collins.

Barthes, Roland. 1975. *S/Z*, trans. Richard Miller. London: Jonathan Cape.

Bernstein, Charles. 1997. "What's Art Got to Do with It? The Status of the Subject of the Humanities in an Age of Cultural Studies." In *Beauty and the Critic. Aesthetics in an Age of Cultural Studies*, ed. James Soderholm, 41–45. Tuscaloosa and London: University of Alabama Press.

Bolter, Jay David. 1991. *Writing Space: The Computer, Hypertext, and the History of Writing*. Hillsdale, NJ: Lawrence Erlbaum.

Bolter, Jay David, and Richard Grusin. 1999. *Remediation: Understanding New Media*. Cambridge: MIT Press.

Burwick, Frederick, ed. 1999. Lessing's Laokoon. *Context and Reception. Poetics Today* 20, no. 2 (special issue).

Christin, Anne-Marie. 1995. *L'image écrite* [The Written Image]. Paris: Flammarion.

Christin, Anne-Marie. 1999. *Vues de Kyoto* [Views of Kyoto]. Lectoure, France: Éditions du Capucin.

Christin, Anne-Marie. 2000. *Poétique du blanc* [A Poetics of the White Space]. Leuven, Belgium: Éditions Peeters and Vrin.

Derrida, Jacques. 1974. *Glas* [Bell]. Paris: Galilée. (English trans. by John P. Leavey, Jr., and Richard Rand. Lincoln: University of Nebraska Press, 1986).

Evans, Jessica, and Stuart Hall. 1999. "Introduction." In *Visual Culture: The Reader*, ed. Jessica Evans and Stuart Hall, 1–7. London: Sage.

Genette, Gérard. 1987. *Seuils* [Thresholds]. Paris: Éditions du Seuil.

Heffernan, James. 1995. *Museum of Words.* Chicago: University of Chicago Press.

Kirschenbaum, Matthew. 1999. "Media, Genealogy, History." In *Electronic Book Review* 9. Available: <http://www.altx.com/ebr/reviews/rev9/r9kir.htm>.

Lakoff, George. 1984. *Women, Fire, and Dangerous Things.* Chicago: University of Chicago Press.

Landow, George. 1997. *Hypertext 2.0*, Baltimore: Johns Hopkins University Press.

Langacker, Ronald. 1987. *Foundations of Cognitive Grammar.* Stanford: Stanford University Press.

Lapham, Lewis H. 1994. "Introduction to the MIT Press Edition." In Marshall McLuhan, *Understanding Media: The Extension of Man*, i–xxiii. Cambridge: MIT Press.

Latour, Bruno. 1987. *Science in Action.* Cambridge: Harvard University Press.

Lynch, Michael. 1991. "Science in the Age of Mechanical Reproduction: Moral and Epistemic Relations between Diagrams and Photographs." *Biology and Philosophy* 6:205–226.

Lynch, Patrick J., and Sarah Horton. 1999. *Web Style Guide: Basic Design Principles For Creating Web Sites.* New Haven: Yale University Press.

McLuhan, Marshall. [1964] 1994. *Understanding Media: The Extension of Man.* Cambridge: MIT Press.

McLuhan, Marshall, and Quentin Fiore. [1967] 1996a. *The Medium Is the Massage.* San Francisco: Hardwired.

McLuhan, Marshall, and Quentin Fiore. [1968] 1996b. *War and Peace in the Global Village.* San Francisco: Hardwired.

Mélon, Marc. 1996. "Une très vieille et vague cousine de Bretagne? Roman-photo et photo-essai." [Some old and barely known cousin from nowhere? Photonovella and photographic essay.] In *Le roman-photo*, ed. Jan Baetens and Ana Gonzalez, 138–157. Amsterdam: Rodopi.

Messaris, Paul. 1994. *Visual Literacy. Image, Mind & Reality.* Boulder, CO: Westview.

Messaris, Paul. 1997. *Visual Persuasion.* London: Sage.

Meunier, Jean-Pierre. 1998. "Connaître par l'image" [To know by the Image]. *Recherches en communication* 10:35–75.

Mirzoeff, Nicholas. 1998. "What Is Visual Culture?" In *The Visual Culture Reader*, ed. Nicholas Mirzoeff, 1–13. London and New York: Routledge.

Mitchell, W. J. T. 1985. *Iconology*. Chicago: University of Chicago Press.

Mitchell, W. J. T. 1994. *Picture Theory*. Chicago: University of Chicago Press.

Mitchell, William J. 1992. *The Reconfigured Eye*. Cambridge: MIT Press.

Nunberg, Geoffrey ed. 1998. *The Future of the Book*. Berkeley and Los Angeles: University of California Press.

Osman, Colin. 1987. "Photography Sure of Itself." In *A History of Photography*, ed. Jean-Claude Lemagny and André Rouillé, 165–185. Cambridge: Cambridge University Press.

Plissart, Marie-Françoise. 1985. *Droit de regards (avec une lecture de Jacques Derrida)* (Right of Insight [with a Lecture by Jacques Derrida]). Paris: Éditions de Minuit.

Plissart, Marie-Françoise. 1998. *Right of Inspection (with a Lecture by Jacques Derrida)*. New York: Monacelli.

Ribière, Mireille, ed. 1995. *Photo Narrative. History of photography* 19, no. 4 (special issue).

Tufte, Edward. 1990. *Envisioning Information*. Cheshire, CT: Graphics Press.

Williams, Raymond. [1974] 1990. *Television: Technology and Cultural Form*, ed. Ederyn Williams. London: Routledge.

| 9 |

COGNITIVE AND EDUCATIONAL IMPLICATIONS OF VISUALLY RICH MEDIA: IMAGES AND IMAGINATION

Jennifer Wiley

As images and animations become easier to use within electronic texts, the nature of "written communication" is changing. A fundamental question, as such communications can no longer be assumed to be composed primarily of words, is, how does this evolution in the nature of written communication change the transmission of knowledge? Surely images are a powerful means of presentation, allowing for vivid and indelible experiences on the part of the reader, and in some cases a picture may be worth a thousand words. However, although under some conditions images may allow for an immediate apprehension of a new concept, research in cognitive science has demonstrated that in other cases, images and animations may actually lead to poorer understanding and distract the reader from understanding the central message of a text. In an age in which visual forms of presentation are becoming more and more prevalent, it is important for authors and educators to recognize the potential cognitive implications of including graphics, pictures, or animations alongside, or instead of, prose.

In the past, Arthur Glenberg and Mark McDaniel (1985), among other cognitive researchers, have argued whether text alone can be an adequate surrogate for experience. A related question is whether visuals can be an adequate surrogate for educational text. An important aspect of text and the way a text conveys information is its structure. Although in some ways the structure of a linear text confines a reader to a set path, it is also a very powerful tool for communication, as it leads a reader through inferences and elaborations to the conclusions of an argument only after important and relevant evidence or premises have been advanced. Well-written educational text guides the reader to new understanding, not just through the transmission of information, but through allowing the reader to build a conceptual model of the subject matter. To what extent can images change this relationship with text? This chapter will examine the educational implications of visual adjuncts and how they may affect the processing of conceptual information and therefore, the transmittal of knowledge within particular subject matter areas. In short, this chapter will examine advantages that can be obtained from including images in text (as adjuncts to the textual presentation of information), but it will also note a number of conditions that constrain what kinds of images are useful and how they should be presented. Further, instances in which presenting visuals can interfere in the relationship with text and actually prevent readers from developing understanding will also be investigated.

Specifically, evidence from studies in cognitive science will be brought to bear on the question: How does the ability to embed images and animations into multimedia webs of information affect our ability to convey information that has been traditionally transmitted via written text? Note that the emphasis in this chapter will be on cognitive processes that underlie and relate to understanding: the comprehension of a text, conceptual memory for the new information, and the ability to use new information in decision making and problem solving. The first part of this chapter will give the reader a background in studies that have shown the cognitive benefits and costs of using multimedia presentation in relation to learning in specific subject matter areas. The latter part of the chapter discusses when and how images and text should be presented to maximize the benefits and minimize the potential cognitive costs of visually rich presentation.

What Are the Benefits of Visually Rich Presentation?

One thing that visually rich presentation surely provides for the student is an image. And images have several advantages over text. First, as Diane Schallert (1980) found, students prefer images or texts with images (whether pictures, graphics, or animations); readers will choose illustrated text over plain text most of the time. Images, especially animated images and virtual simulations, can make the learning experience closer to real-world experience, and as a consequence the experience can be more vivid and more engaging for the learner as has been demonstrated by Meredith Bricken and Chris Byrne (1993) and the work of Christopher Dede and his colleagues (Dede 1995; Dede et al. 1999). Further, K. Ann Renninger, Suzanne Hidi, and Andreas Krapp (1992) have found that readers are more interested in illustrated text and tend to spend more time on readings in which they are interested, which has been shown to improve learning in some cases.

Bernd Weidenmann (1989) makes the important point that the way that we process images is different than the way we process text. Whereas reading text requires readers to fixate on every word or two, for around 300 milliseconds a word, going from left to right and from the top line to the bottom line, images are scanned much more quickly and globally. Processing of images is much more "holistic," as people generally assume that a cursory scan across an image is sufficient for them to process the content of the image. Thus, people may feel that images are easier to process and require less effort than words do. There is some evidence (for example, in Larkin and Simon 1987) that more information can be considered at once in graphic form and that it is easier to shift one's attention among different parts of an image than among different features mentioned in text. Thus, there is some support for the idea that image-related processing is less effortful. Although scanning a picture takes less effort than reading, however, several studies by Susan Palmiter and Jay Elkerton (1983), and Stephen Payne, Louise Chesworth and Elaine Hill (1992) have found that it also generally yields only superficial processing of information. Conditions that lead to better comprehension tend to require effort, including refixations and deep thought about features and relations, whether in an image or an animation or in a text. Similarly, although images may be more attractive, in that they are more pleasing, they are also more distracting, in that they attract attention even when the reader would be better off engaged in deep thought or reading. Beyond these limitations, which will be discussed in more detail later in the chapter, there are some very robust reasons why image-based presentation can sometimes be a quite effective way to convey new information.

One example of when images are beneficial is situations in which deep processing of all presented information is simply not possible, especially when there is a massive amount of information that is available, as in the case of information streams from space stations or weather satellites. As Lloyd Rieber (1995) suggests, in cases when the amount of information to be processed is overwhelmingly huge and from many possible dimensions, the simplification provided by converting the multidimensional data into a dynamic visual representation may be the only way to get a handle on the incoming stream of data. Some studies suggest that images provide a more efficient way of presenting multiple dimensions and the relationships between those dimensions than text (e.g., McDaniel and Waddill 1994), and Dede et al. (1999) note that this may be the case especially to the extent that designers use preattentive cues such as color and motion to direct the viewers' attention to relevant features of the image.

Another possible advantage of visually rich presentation is that it may allow for information to be represented in memory in multiple ways. Alan Paivio's dual-coding theory (Paivio 1986; Clark and Paivio 1991) suggests that there are two forms in which information can be represented in our memories. One of these forms is predominantly verbal, and the other is visuospatial. A great deal of previous research within cognitive psychology has suggested that, to the extent that information can be coded redundantly (that is, in both verbal and visual form), we are more likely to remember that information. Presenting images along with text can improve our memory for the concepts that are presented, as we will have dual traces (or two forms) of memory for the ideas mentioned in the text. Thus, to the extent that images allow for multiple forms of representation in memory, visually rich presentation may improve our memory (that is, our recall) of text.

On another level, images may improve our comprehension of text. In many cases the ability to visualize a situation is related to our understanding of the information that has been presented. "Seeing" the path of Stephen Daedelus through Dublin or the circular chain of actions that corresponds to photosynthesis and the Krebs cycle means that we have achieved a certain level of comprehension in reading from James Joyce or a high school biology text. Cognitive psychologists, including Deirdre Gentner and Albert Stevens (1983), have referred to this ability to visualize a process as a "runnable mental model." Further, it would seem that an expedient way of enabling a student to visualize the process of photosynthesis and to attain a mental model of the process would be to show him or her a visual simulation or animation of the process.

Similar to this concept of a mental model, Walter Kintsch's (1998) construction integration model posits that readers build a similar structure, a situation model, as they try to comprehend text. In the cognitive literature, it is assumed that as readers process text, they attempt to represent it in their memory. One form of representation is relatively close to direct perceptual experience and can be seen as a "surface" model. In the case of reading text, the exact words that were used would be a part of this surface model. On the other hand, as a reader attempts to understand the meaning of the input, she develops a situation model, which is an abstraction of the text. The situation model includes the gist or meaning of the new input, integrates the new information with information in memory, and represents the reader's causal or conceptual understanding of the subject matter. In many cases, whether it is a representation of the action sequences in a novel or an understanding of how photosynthesis works, this model can be image based. In such cases, situation models and mental models would be quite similar, and as Wolfgang Schnotz, Justus Bockheler, and Harriet Grzondziel (1999) suggest, it is easy to see how a simulation or image would directly facilitate the creation of such a model. Even when the understanding of the subject matter does not necessarily require an image-based representation, however, an image can still facilitate the creation of a situation model if it provides the basis for an abstract model of the content of the text. Hence, figures, graphs, or flowcharts that may enable the reader to think about abstract concepts through images may allow for the creation of more complete situation models and as a consequence may in fact improve comprehension of text, as the studies of Darrell Butler (1993) and William Winn (1988) have demonstrated.

Finally, it is possible that in some cases, images alone without text may be the best way to promote understanding of some concepts and attack some problems. Work in verbal overshadowing by Jonathan Schooler, Stellan Ohlsson, and Kevin Brooks (1993) has shown that in some cases images can be overpowered by text or labels. Perhaps the best example is in insight problem solving, in which representing a problem verbally sometimes limits the discovery of possible solutions. Anecdotes of discovery are rife with examples in which imagery played an important role; for example, Amit Subhash Kulkarni's discovery of the benzene ring is said to have been inspired by a vision of a snake chasing its tail. Here the experience in a visual space in which the problem solver can manipulate the problem without the limitations of language is thought to contribute to true discovery. A second arena in which the image may be superior to language are situations in which experts from two different fields are attempting to collaborate on a problem. Marek Kohn (1994) notes that using an image to convey a problem can help experts without a common vocabulary talk across disciplines. It is unclear how much of everyday learning may be able to take place in such a linguistic vacuum, but there may be instances in which an image needs to stand alone, so that each viewer of the image is free to develop an understanding on his own terms, whether using his own vocabulary or in a nonverbal way. Particular domains—those in which information is inherently spatial or complex—may lend themselves to visual presentation of information. Similarly, work by Timothy Hays (1996) on individual differences in learner preferences suggests that some learners may require images to attain an understanding of subject matter, whereas other students may prefer a more verbal mode of communication. All of these factors will affect a particular student's ability to learn from visually rich presentations.

When Visually Rich Presentation Fails

With all these potential benefits of images, it is perhaps surprising that for the past forty years, the empirical results of studies on learning from text with visual adjuncts have been less than positive. In an early review of studies using illustrated texts, S. Jay Samuels (1970) found little support for the superiority of illustrated text over plain text. In fact, in some cases, Samuels found, illustration leads to poorer learning than simple text presentation. Follow-up investigations suggest that one reason for the lack of a consistent positive effect of images on learning is that any learning effect depends greatly on the kind of image that is used. In a review by Joel Levin, Gary Anglin, and Russell Carney (1987) that discriminated between decorative illustrations and representative, organizing, and interpretive images, decorative illustrations, not surprisingly, were found to lead to the smallest improvements in, and sometimes even to negative effects on, learning. Decorative illustrations are often not relevant for the important concept that is conveyed by the text, yet they are still interesting for the reader and will attract the reader's attention. For this reason, emotionally interesting but irrelevant pictures can be seen as part of a larger class of "seductive-details" effects that have been studied by Ruth Garner, Mark Gillingham, and Stephen White (1989) and Shannon Harp and Richard Mayer (1998). Seductive details are elements of a text that do not support a main point but to which readers choose to devote a large portion of their attention. Photographs containing interesting scenes and pictures of people are often emotionally interesting. Yet, especially in the understanding of science text, Harp and Mayer found that they generally distract the reader, focusing her on irrelevant prior knowledge, thus leading to poorer understanding of the content. Similarly, Anne Treisman and G. Gelade (1980) suggest that color and motion are preattentive cues that necessarily attract a reader's focus. If they are not used to emphasize conceptually important information, they too can seduce the reader. Thus, conceptually irrelevant visuals do not help communication, and in some ways may impede it.

A second caveat for using visually rich presentation is that even when images are relevant for understanding the target concept, there is a danger that images or animations may make knowledge acquisition too easy or effortless. Weidenmann (1989) notes that people tend to feel that a short glimpse of an image is generally sufficient for understanding. This can lead to an illusion that they understand a graphic or image and that they have absorbed all the information that is available in it even when they have not really engaged in deep thought about the information. On a second level, sometimes an image, especially an animation, can provide so much information so easily that readers will have a good idea of how a dynamic system works (for example, they may know what the process of photosynthesis looks like in action) but they may fail to develop a good understanding of why the system works the way it does (so that they could re-create the system or apply their knowledge to a new instance, as has been shown in Palmiter and Elkerton 1983 and Payne, Chesworth, and Hill 1992. It may be that such understanding can come only from having to read text and imagine the dynamic system themselves. Although in such cases the visualizations are in fact communicating the descriptive information quite effectively, students may not actually be learning to the same extent as they would if they were able and compelled to generate the visualization themselves. This effect has in fact been demonstrated in studies by Linda Gambrell and Paula Jawitz (1993), and Schnotz, Bockheler, and Grzondziel (1999), in which the act of imagining leads to better learning than when the complete mental models are provided for students. In a number of instances, still pictures or still sequences of pictures, from which readers needs to infer movement for themselves, have led to better understanding of dynamic systems than animations that actually show the motions (e.g., Hegarty et al. 1999). Further, the studies by Palmiter and Elkerton and Payne, Chesworth, and Hill suggest that this active construction of a runnable mental model is especially important for long-term learning.

Of course this "constructivist" approach is limited by several variables, such as the reader's level of knowledge and ability. Readers can use their imagination to visualize a scene only when they know what to imagine, so images that provide readers with the basis of a mental model and animations that show the dynamics of a model may be especially important for people who lack such knowledge. The goal in instruction is to find a match between readers' knowledge and the amount of visualization that they can accomplish on their own, so that they can engage in a maximal amount of active processing of information, with a reasonable probability that they will be creating a correct mental model of the subject matter. The bottom line is that in some respects the goal of educational text is not simply the transmission of information, but the production of appropriate mental representations of the information on the part of the learner. The goal of the learner is not necessarily acquiring a correct descriptive visualization of the phenomenon being explored, but the construction of a correct conceptual or causal explanatory model, which may require some effort on the part of the learner.

One might ask whether there is a parallel in aesthetic experience. Simply having a "mental model" of the path that a novel's main character takes through a town does not really give us the experience of the main character. Instead, reading about the character's path through the town conjures up particular feelings and insights and gives us a sense of immersion in the episode as we experience the revelations and epiphanies of the characters. Although any reader may know from the Cliff's Notes version where the main character ends up and how he got there, there is a sense of intimacy and intensity that comes from re-creating the novel in one's own mind's eye. Similarly in learning, true understanding may not be in knowing the end state of a problem, but in the experience of constructing an explanatory construct that leads one to the answer.

A final aspect of images that affects their communicativeness, especially within educational texts, is the extent to which they are realistic images versus symbolic or abstract illustrations. Although images that are highly realistic may be easier to process, and perhaps more engaging, they are less likely to convey conceptual knowledge to the viewer. Animations that are reproductions of real-world actions are more effective if they are "doctored" to emphasize important features of the display (e.g., Faraday and Sutcliffe 1997). For instance, Mary Hegarty et al. (1999) found that animations with labels, arrows, or narrations that emphasize important features, relations, and movements lead to better learning outcomes than animations that are merely real-time simulations. And, Hari Narayanan and Hegarty (1998) found that animations that are stoppable and restartable under the learner's control may lead to better learning than real-time simulations. Similarly, images that represent systems can use either representations that are very faithful to real life or more abstracted representations. For instance, Judith Effken, Namgyoon Kim, and Robert Shaw (1997), investigated the effects of image type on a lesson on the circulatory system. Students saw either images of a human heart and blood vessels or a schematic model of the circulatory system. The schematic abstraction may require more effort to process, but the effort is directly relevant to representing important relations and elements in memory. Hence, building on the principle that active processing of information is better, especially when students are forced to construct their own mental model of the information, it would follow that abstract models may be more helpful in acquiring an understanding of the circulatory system than more real-life images. In Effken, Kim, and Shaw's study, the schematic diagram supported better decision making and understanding of the consequences of drugs on the circulatory system than the more realistic display. Again, individual differences among learners will affect whether and the degree to which symbolic representations enhance learning, and perhaps only students of higher ability will be able to benefit from symbolic representations, as the process of "unpacking" a diagram or figure may impose a load on processing. To the extent that students can handle the extra demands, however, images with more symbolic representations seem to focus their attention on conceptually important features, which can result in better understanding.

Based on the empirical studies discussed above, it seems that visuals that are faithful to real-world experiences in many ways are not as beneficial for learning as more abstract and incomplete sources of visual information. In short, including images in text can have advantages, but there are also a number of conditions that constrain what kinds of images are useful and how they should be presented. There are also instances in which presenting visuals can interfere in the relationship with text and actually prevent readers from developing understanding.

When and How to Use Visual Adjuncts

In a series of studies, I have been examining which conditions lead to the best understanding from electronic multimedia. So, for example, I have manipulated features of Web sites such as the placement and kind of overview that readers may get, as well as the number of windows they are given in which to read documents. In general, the design of my Web experiments is to have students read through a Web site at their own pace. The purpose of reading through the documents is usually to write an opinion-based essay on a topic such as "What caused the significant change in Ireland's population between 1846 and 1850?" or "What caused the explosion of Mt. St. Helens?" or "What impact did the building of Grand Coulee Dam have on Washington state?" Following the writing of the essay, students are presented with several questions to assess their understanding of the material they have read. Included in the test questions are some factual recall items, as well as items that require conceptual understanding or inferences based on the presented material. Learning is assessed by examining both the essays that students write and their performance on the posttests as well. As a result, the effects of different kinds of presentation of information can be assessed.

The first principle of effective visual presentation that I have been investigating is minimizing the possibility of images' acting as seductive details. There are two ways of achieving this goal: minimizing the possibility of competition between picture and text and ensuring that the pictures used are relevant to the material presented. In relation to the first alternative, if one chooses to use certain pictures or images whose chief attraction to viewers is their emotional interest, they should be presented in a way that does not compete with text, for example, on a separate page in a Web site. In line with this theory, in one experiment students were presented with photographs showing the aftereffects of earthquakes (e.g., collapsed bridges and cars sticking out of ditches) either at the same time as a text about the causes of earthquakes (in the same document, as on a standard textbook page), or immediately before the text, or in hyperlinks from the text in which the pictures were presented in their own window.

When pictures were presented at the same time as the text, there was a clear seductive-details effect; that is, students were less likely to develop an understanding of what causes earthquakes when they had pictures embedded in the text than were students who read plain text. When students had access to the photographs through a link, they also showed a seduction effect and learned as little from the text as the embedded-pictures group had. Students who received the emotionally interesting pictures beforehand, however—and could not return to them while they were reading text—did not show a seduction effect. They had the same amount of learning as students with no pictures but rated the task as more interesting. This suggests that even emotionally interesting images can be used to pique interest in scientific subject matter, as long as they are presented in a way that does not directly detract from the amount of time spent reading a text containing conceptual information. Patricia Wright, Robert Milroy, and Ann Lickorish (1999) and Rieber (1992) found similar results when they looked at learning from animated graphics. Students learned better from animated graphics when they were presented separately from text.

A further twist on the use of evocative images in text is that in some domains, emotionally interesting pictures, for instance, photographs of people, may not be just seductive details. In particular, history teachers often use primary sources such as photographs to prompt students to think about specific historical contexts. It is possible that in such contexts, emotionally interesting adjuncts such as photographs may improve rather than detract from comprehension. Following this intuition, in the same study described above, a second experiment used historical subject matter (based on a Web site about Columbia River history). As in the study using texts about earthquakes, students were presented with either plain text, text with links to images, or text with images embedded in the text pages. Whereas emotionally interesting photographs embedded in text seemed to distract readers from the scientific text on earthquakes, the embedded photographs and images on the history site made readers of the history text more interested, led to more time spent on the reading task, and improved understanding on tests of the subject matter following the reading phase.

There may be many factors underlying this difference in learning from illustrated text in history and science. One possible explanation is that the texts were not matched for difficulty, and it is possible that the concepts in the science texts were more difficult, and thus more vulnerable to competition from the images. A second possible explanation is that in order to understand how earthquakes happen, readers may need to create a visual model, and the presence of irrelevant pictures may have directly interfered with the visual processing required to create such a model. On the other hand, understanding how Native Americans were displaced and the effects of the Grand Coulee Dam on the Columbia River region may not have required any visual processing, in which case understanding was not in competition with the presented images. A third possibility is that emotionally interesting photographs of people do motivate readers to learn more about the people they see, which may in fact prompt them to develop a better understanding of the historical context. For whichever reason, it is interesting that a seductive-details effect was avoided and the "emotional" images did not have negative effects on learning from historical text.

As noted earlier, the second way that one may avoid seduction effects is simply to make sure that the images that are presented are relevant for the concepts to be learned. In this way, any time that is spent viewing a particular image should actually be helping the reader to better understand the concepts that are presented in the text. A corollary to this principle is that images will be especially communicative to the extent that they highlight important conceptual relations and emphasize key features either through salience or symbolism. Similarly, images may be particularly helpful toward learning when they have structure that makes important conceptual relations obvious or when they provide organization for new knowledge, especially knowledge that is abstract.

Another principle of visually rich presentation is that when one wants readers to learn from a text, it is best to present them with just enough information to enable them to construct and imagine their own visualizations of the text. That said, there are individual differences in spatial ability and visualization skill that make images and animations more important for understanding for some readers. In one study of individual differences, Hays (1996) found that readers with low spatial ability achieved a better understanding of the chemical principle of diffusion after viewing an animation, but other studies have shown that students who have good visualization skill are generally better off generating their own visualization of dynamic processes from still pictures.

Finally, when subject matter is complex and dynamic, such as that involving interpreting streams of data from weather satellites or space stations, or just difficult to visualize, then visuals and animations may help people understand the subject matter regardless of ability. For example, Vicki Williamson and Michael Abraham (1995) found that animations helped students understand the particulate nature of matter, a concept from chemistry of which it is difficult for students to form an accurate image. Similarly, visual representations of complex data can give human thinkers the ability to consider many more dimensions, and the salient relationships between those dimensions, than they might otherwise.

The fact that images are most effective toward communication when they are conceptually relevant may seem obvious, but open almost any textbook and you will find scores of images that serve purely decorative purposes. In fact, in a recent survey of publications, Butler (1993) found an increase in the percentage of pictures included in text but *not* in the percentage of data representations or conceptual graphics. Thus, as electronic publishing makes visually rich presentation even easier, the trend toward inclusion of greater numbers of decorative images may continue and seep into new media. In particular, the capacity of the CD has encouraged a "kitchen sink" approach to instructional support. Load up any of the CD-ROMs that publishers are now including with textbooks and look at the wide range of images that are available and the range of relevance they have to the subject matter of the textbooks they accompany. Although such image collections may lend themselves to romantic notions of enabling limitless and unconfined creative exploration, just as with hypertext, effective knowledge acquisition usually requires structure, coherence, and relevance. Browsing and surfing through images may allow the viewer to make some novel associations, but we must remember that "chance favors the prepared mind" and for most students, a battery of images of varying degrees of relevance will only overwhelm them and obscure any lesson that is intended to be learned from the images.

The thrill that one anticipates when one envisions the facilitation of new connections in students' minds through exposure to imagery is a siren's call to inclusion of more images in the student experience, but extensive gains in this area are not realistic in everyday practice. We must remember that there is also a "joy of discovery" that comes from reading carefully crafted text and epiphanies that a reader may experience that are engineered by the author. Although imposing a linear structure or limiting the presentation of images to specific windows of opportunity necessarily reduces the chances of remote, novel, or random discoveries that a reader might make, on the other hand, it supports a larger number of smaller inferences that will support a better, more basic understanding of the topic in most readers. This is one reason why a return to a pictorial age seems unlikely in education. The construction of an argument that guides the student through evidence to a new conclusion or conceptual understanding is still perhaps the most reliable means for transmitting subject matter knowledge. Visual adjuncts can serve an important role in clarifying and providing vivid examples of evidence and in exciting the reader about a topic, perhaps even in providing an aesthetic or persuasive experience, but the images and animations themselves can hardly stand alone in terms of subject matter learning.

Perhaps the most striking evidence for this conclusion are recent results from experiments with virtual reality. Virtual reality, in which the viewer has direct experience with images, is the ultimate visualization tool. Here, the "reader" is now an "experiencer," and the potential exists to convey an understanding of new concepts through direct experience. Yet in a review of the literature on virtual reality, Joshua Hemmerich and I found that virtual reality experiences are not easily translated into learning experiences, and the results of studies on the educational uses of virtual reality underscore the same principles as have been discussed in this chapter with respect to the use of other visual forms (Wiley and Hemmerich forthcoming). The bottom line is that students do not learn by the simple transmission of information. Virtual reality adds value to educational contexts when it goes beyond "realistic" experiences in ways that make elements important for conceptual understanding obvious, while leaving some work in constructing a coherent representation up to the student.

The way that information is transmitted can have a significant influence on understanding, and any change in the modal form of communication is sure to have an impact on educational practice. Just because a dynamic image-based representation and the ability to visualize a particular scene or process is a desired end state, however, does not mean that a visually rich presentation with great fidelity to the real-world situation is going to be the best medium for educational communication. At least at this point, what we can conclude from the evidence in the cognitive literature is that students need structure and emphasis on important elements to develop understanding, whether they are learning from text or images. That said, images can play an important role in supporting written texts, as well as making them more memorable and vivid for the reader, as long as they are conceptually relevant, and as long as something is still left to the imagination.

Acknowledgments

The preparation of this chapter was supported by grants to the author from the Paul G. Allen Virtual Education Foundation and the Office of Naval Research, Cognitive and Neural Science and Technology Program. The opinions expressed in this chapter should not be taken to reflect the positions of these organizations.

Works Cited

Bricken, Meredith, and Chris Byrne. 1993. "Summer Students in Virtual Reality." In *Virtual Reality: Applications and Exploration*, ed. A. Wexelbalt, 199–218. New York: Academic Press.

Butler, Darrell. 1993. "Graphics in Psychology." *Behavior, Research Methods, Instruments & Computers* 25:81–92.

Clark, James, and Alan Paivio. 1991. "Dual Coding Theory and Education." *Educational Psychology Review* 3:149–210.

Dede, Christopher. 1995. "The Evolution of Constructivist Learning Environments." *Educational Technology* 35:46–52

Dede, Christopher, Marilyn C. Salzman, R. Bowen Loftin, and Debra Sprague. 1999. "Multisensory Immersion as a Modeling Environment for Learning Complex Scientific Concepts." In *Computer Modeling and Simulation in Science Education*, ed. N. Roberts, W. Feurzeig, and B. Hunter, 282–319. New York: Springer-Verlag.

Effken, Judith, Namgyoon Kim, and Robert Shaw. 1997. "Making the Constraints Visible." *Ergonomics* 40:1–27.

Faraday, Peter, and Alastair Sutcliffe. 1997. "An Empirical Study of Attending to and Comprehending Multimedia Presentations." In *Proceedings ACM Multimedia 96 Conference*, 265–275. New York: ACM Press.

Gambrell, Linda, and Paula Jawitz. 1993. "Mental Imagery, Text, Illustrations and Children's Story Comprehension and Recall." *Reading Research Quarterly* 28:264–276.

Garner, Ruth, Mark Gillingham, and C. Stephen White. 1989. "Effects of 'Seductive Details' on Macroprocessing and Microprocessing in Adults and Children." *Cognition and Instruction* 6:41–57.

Gentner, Deirdre, and Albert Stevens. 1983. *Mental Models.* Hillsdale, NJ: Lawrence Erlbaum.

Glenberg, Arthur, and Mark McDaniel. 1985. "Mental Models, Pictures and Text." *Memory and Cognition* 20:458–460.

Harp, Shannon, and Richard Mayer. 1998. "How Seductive Details Do Their Damage." *Journal of Educational Psychology* 90:414–434.

Hays, Timothy. 1996. "Spatial Ability and the Effects of Computer Animation on Short-Term and Long-Term Comprehension." *Journal of Educational Computing Research* 14:139–155.

Hegarty, Mary, Jill Quilici, N. Hari Narayanan, Selma Holmquist, and Roxana Moreno. 1999. "Multimedia Instruction: Lessons from Evaluation of a Theory-Based Design." *Journal of Educational Multimedia and Hypermedia* 8:119–150.

Kintsch, Walter. 1998. *Sources of Comprehension.* Cambridge: Oxford University Press.

Kohn, Marek. 1994. "Is This the End of Abstract Thought?" *New Scientist*, September 17, pp. 37–39.

Larkin, Jill, and Herbert Simon. 1987. "Why a Diagram Is (Sometimes) Worth Ten Thousand Words." *Cognitive Science* 11:65–100.

Levin, Joel, Gary Anglin, and Russell Carney. 1987. "On Empirically Validating the Functions of Pictures in Prose." In *The Psychology of Illustration*, ed. D. Willows and H. Houghton, 51–85. New York: Springer-Verlag.

McDaniel, Mark, and Paula Waddill. 1994. "The Mnemonic Benefit of Pictures in Text: Selective Enrichment for Differentially Skilled Readers." In *Comprehension of Graphics*, ed. W. Schnotz and R. Kulhavy, 165–181. Amsterdam: Elsevier.

Narayanan, N. Hari, and Mary Hegarty. 1998. "On Designing Comprehensible Interactive Hypermedia Manuals." *International Journal of Human-Computer Studies* 48:267–307.

Paivio, Alan. 1986. *Mental Representations: A Dual Coding Approach.* New York: Oxford University Press.

Palmiter, Susan and Jay Elkerton. 1983. "Animated Demonstrations for Procedural Compter-Based Tasks." *Human Computer Interaction*, 8:193–216.

Payne, Stephen, Louise Chesworth, and Elaine Hill. 1992. "Animated Demonstration for Exploratory Learning." *Interacting with Computers* 4:3–22.

Renninger, K. Ann, Suzanne Hidi, and Andreas Krapp. 1992. *The Role of Interest in Learning and Development.* Hillsdale, NJ: Lawrence Erlbaum.

Rieber, Lloyd. 1992. "Using Computer Animated Graphics with Science Instruction with Children." *Journal of Educational Psychology* 82:135–140.

Rieber, Lloyd. 1995. "Visualization as an Aaid to Problem Solving: Examples from History." *Educational Technology Research and Development* 43: 45–56.

Samuels, S. Jay. 1970. "Effects of Pictures on Learning to Read, Comprehension, and Attitudes." *Review of Educational Research* 40:397–407.

Schallert, Diane. 1980. "The Role of Illustrations in Reading Comprehension." In *Theoretical Issues in Reading Comprehension*, ed. Rand J. Spiro, B. C. Bruce, & William F. Brewer, 503–525. Hillsdale, NJ: Lawrence Erlbaum.

Schnotz, Wolfgang, Justus Bockheler, and Harriet Grzondziel. 1999. "Individual and Cooperative Learning with Interactive Animated Pictures." *European Journal of Psychology of Education* 14:245–265.

Schooler, Jonathan, Stellan Ohlsson, and Kevin Brooks. 1993. "Thoughts beyond Words." *Journal of Experimental Psychology: General* 122:166–183.

Treisman, Anne, and G. Gelade. 1980. "A Feature-Integration Theory of Attention." *Cognitive Psychology* 12:97–132.

Weidenmann, Bernd. 1989. "When Good Pictures Fail." In *Knowledge Acquisition from Text and Pictures*, ed. H. Mandl and J. Levin, 157–171. New York: Elsevier.

Wiley, Jennifer, and Joshua Hemmerich. Forthcoming. "Literacy: Learning from Multimedia Sources." In *Encyclopedia of Education.* New York: Macmillan Library Reference.

Williamson, Vicki, and Michael Abraham. 1995. "The Effects of Computer Animation on the Particulate Mental Models of College Chemistry Students." *Journal of Research in Science Teaching* 32:521–534.

Winn, William. 1988. "Recall of the Pattern, Sequence, and Names of Concepts Presented in Instructional Diagrams." *Journal of Research in Science Teaching* 25:375–386.

Wright, Patricia, Robert Milroy, and Ann Lickorish. 1999. "Static and Animated Graphics in Learning from Interactive Texts." *European Journal of Psychology of Education* 14: 203–224.

IV

Identities and Cultures in Digital Designs

10

FEMINIST CYBORGS LIVE ON THE WORLD WIDE WEB: INTERNATIONAL AND NOT SO INTERNATIONAL CONTEXTS

Gail E. Hawisher and Patricia Sullivan

For the past few years, the research in which we've been engaged has examined the on-line lives of women. Most recently this research has taken the form of looking at how women, in and out of academe, represent themselves visually on the World Wide Web and how they are represented on the Web by others. We began this research to address our concern that the dominant societal norms of patriarchy would be the primary force shaping the Web and thus would tend to represent women only as they have traditionally been depicted. Here we'd like to bring together our studies and add an international component to explore how women—both here and in other countries—forge identities for themselves visually and verbally in on-line contexts.

Our thinking in this chapter draws on at least two strands of feminist thinking that can be brought to bear on investigations of gender in online identity: feminist technology theory and cultural geography. Feminist technology theory in writing studies contributes key concerns to our study of Web identities: how subjectivity and agency operate for women on the Web and the importance of an active presence on the Web. Though many disciplines have been interested in how the Web body can enact post-modern notions of fragmented subjectivities, feminists in writing and technology studies have focused primarily on the construction of agency and subjectivity in cyberspace. Intrigued by the concept of cyborgs, feminist writing researchers, such as Cynthia Selfe 1992, Laura Sullivan (1997), Sarah Sloane (1999), and Joanne Addison and Susan Hilligoss (1999) have interrogated how Internet and Web subjectivities are partial and constructed, depending in part on the particular e-space scrutinized. This approach has allowed for contesting traditional genders, for encouraging gender play, and for experimenting with on-line feminist activism. Feminists in writing studies and technology also have strongly supported the importance of the active presence and participation of women on the Web. Some, such as Dale Spender (1995), have framed this imperative as a feminist economic issue and argue that historically women gain control over production only when a communication medium is no longer in its ascendancy. Thus it becomes critical that women assume a central role as entrepreneurs in Web space. Others, such as Emily Jessup (1991) and H. Jeanie Taylor, Cheris Kramarae, and Maureen Ebben (1993) have framed it as an issue that argues for equitable access for women in all information technology venues. And still others, such as Arturo Escobar (1999), argue for a feminist politics of transformation that alternates between the virtual networks of cyberspace and the physical localities of place to weave the real and the virtual together in ways that link the politics of place and cyberspace.

Feminists in the field of cultural geography have turned their attention to the investigation of geopolitics and postmodern space and offer lenses on how "home" is culturally constructed as a web of place, community, gender, class, ethnic, institutional, disciplinary, and national affiliations. It is not lost on us that the Web consists of home pages to mark spaces that belong to someone or some group, and since home space (often constructed as female) is often contrasted with public space (often constructed as male), the cultural geography of "home" quickly becomes gendered, hierarchical, spatial, and contested. Susan Stanford Friedman (1998) has called this multiply contested space the new geography of identity and has used spatial metaphors to articulate identity as "the site of multiple subject positions, as the intersection of different and often competing cultural formations of race, ethnicity, class, sexuality, religion and national origin, et cetera and so forth" (21). Certainly this complex geography of identity is clearly displayed through the home pages we studied earlier in American settings. Or is it?

In this chapter we review depictions on the Web in a variety of discursive settings—professionals in American academic institutions, college students in the United States, early commercial sites featuring women—and then turn to international feminist sites. The investigation asks whether and how gender on the Web functions as a cultural formation—that is, a formation historically articulated and marked by an accumulation of practices—and how particular cultural practices can come together to construct new gender identities. Our argument has been and continues to be that online Web contexts can provide an opportunity for feminist cyborgs to "act potently" (181) in Donna Haraway's (1991) words: to claim this medium as their own. But we have always included a caveat: The Web does not, indeed cannot, automatically provide this possibility any more than online verbal e-spaces automatically provide egalitarian forums. Without active feminist participation in online spaces, online worlds tend to become controlled and populated by the same forces that have traditionally shaped the colonial subject to (re)produce the very same power structures that electronic spaces are said to disrupt (Selfe 1999).

Earlier Studies

The instinct toward visual representation has been present in our work from the outset (see Hawisher and Sullivan 1998). Several years ago when the two of us began studying the on-line lives of thirty academic women in writing studies, visual authoring on the Web was still in its infancy, and only a few women in the study had created Web pages. As part of that study, however, we discussed on a listserv we named women@waytoofast the many issues women encounter online, including the construction—or not—of images of themselves. The chapter we authored in connection with the study was primarily concerned with verbal online experiences the women encountered. Although we didn't focus on visual dimensions until our experiences on the Web led us to revisit the study from a visual perspective, an emphasis on images was already present in that pre–World Wide Web discussion. As part of the women@waytoofast discussion, some of the women began representing themselves visually and then talked of professional dress and appearance. It all started with one of the women's noting how much trouble she had remembering people online and then writing the following:

```
  xxx
  x x
xxx  x xxxx
  x xx      xx    /
 x          xxxx
x              x
 x         x
  x       x
  xx    xx
    | |
    | |
    / /
```

Well, it wouldn't have to be realistic photos or even faces. Pick an image. I know that this involves even more rhetorical choices, but I know that I'm not the only person in the world who isn't very word-oriented and something visual to hang a message on would help me a lot and others too I think. Why do electronic lists have to be so tyrannically verbal and linear? That's a pretty dorky bird I managed to construct above.

Less than an hour later, another added:

```
   (%%%%)
 (%%%%%%%%)
(%%%   %%%)
 (%%** %%)
  (% : %%)
 (%     %)
```

In Marge Piercy's *He, She, It* an older woman cruises the information networks giving herself a sexy young identity.

Here's mine . The curly hair is accurate at regular intervals when the perms are new.

(I'm not getting older, just better.)

And still another began to theorize about the opportunities online visuals might hold for women:

Re-imaging/imagining ourselves through technology seems like an empowering concept. Very connected to Foucault's technology of the self, gaining access to technology to change how we 'appear' to the world, to revise those Virtual Valerie porn images to something that more accurately represents our lives in the material realm. Besides if Madonna can re-create herself every so often and get dubbed a feminist, why can't we?

Hope everyone has a relaxing holiday.

```
)))/////(((
)))/////(((
))) * *  (((
))) ^    (((
))) o    (((
)))      (((
```

But there were still others who regarded the capability of representing ourselves online very seriously and not without important consequences. Another of the women wrote:

Hi everyone,
The very idea of choosing a face to accompany my online words horrifies me. Should I choose an "authentic" image, one that shows my age and deviations from standard female beauty markers? Or does the electronic medium license me to alter my image? License? Does it *mandate* that I alter my image . . . .

Ah so many rhetorical decisions if we add visual rhetoric. And gender issues become heightened, I think, rather than lessened.

For some, the images foregrounded woman@waytoofast as a safe place in which to play, an e-space where they might risk seeming foolish. Yet they immediately connected the personal and playful aspects of e-space with the serious, professional, and scholarly. E-art was used as a platform for self-representation and for self-critique at the same time that it moved the discussion into areas of how one represents oneself in e-space, in one's department, as a job candidate, or as an untenured faculty member.

Largely as a result of this study and the women's thoughtful responses, we turned from e-art on a list-serv to visual representation on the Web. The Web, with its graphical interface, makes possible the "imaging" or "reimaging" that some of the women on women@waytoofast desired and may indeed allow women to represent one of their many selves more graphically to the rest of the online world. Alan Purves (1998) has written that the coming of print destroyed the importance of the image and made it suspect, but that today the image has returned and is shaping the ways in which the new literate world operates. Gunther Kress (1999) argues that the landscape of the 1990s is "irrefutably multisemiotic" (69) and that "the visual mode in particular has already taken a central position in many regions of this landscape" (69). The Web underscores the importance of the visual to the extent that it firmly folds the visual into communication processes. That is, it allows Web authors simultaneously to construct and to broadcast themselves visually with an ease and speed that hasn't been possible before.

Despite the promise of the Web for visual representation, two years later, when we returned to the thirty women in our original study, several had constructed sites for their departments or programs, but still only a few had experimented with their own home pages. And when we did find home pages, few seemed to be staking out new subject positions; for the most part, the women seemed to be going about academic business as usual, featuring scholarly accomplishments, teaching, and sometimes something about their personal selves. Thus, in our second study (see Hawisher and Sullivan 1999), we extended the analysis to look not only at women's Web constructions within the field of writing studies but also commercial Web pages and professional/personal home pages that featured visual representations of women outside the field. Overall we found that although commercial home pages, such as Victoria's Secret's 1996 Web depictions, did little more than foreground current homogenized representations of "femaleness" and tended to reproduce the age-old stereotypical relations between the sexes, there were clear exceptions to this practice when women themselves took on the task of constructing their own pages. In the Web pages produced by a number of twenty-something women, usually students who had easy access to technology at the time, the women seemed to be staking out multiple subject positions for themselves. They doctored photos, used cartoons, animated quirky representations of themselves, and in general played with the visual in ways that blur the boundaries between physical selves and virtual selves. Cyborg-like, they used technology to capture representations of themselves while at the same time adding and changing bodily features. In displaying, at times, their ears, calves, body piercings, and tattoo-decorated skin, they celebrated their own writings of their bodies. In contrast, the Web representations of faculty women from the earlier women@waytoofast study did little, understandably, to change society's view of academic life, in which women are valued for their minds and knowledge. Their bodies are conventionally thought to be extraneous and, some would argue, potentially damaging to their success in the university setting.

In this second study, then, we began to see how some women in and out of the field of writing studies represented themselves on the Web and how some were often visually constructed by others. Women in the 1996 commercial settings were pictured in on-line advertising in ways that seem familiar to us: as objects to be ogled, objects to stimulate, commodities to be sold and bought. But we also saw examples of the possibilities the new media begin to provide women for activism: for forging new social arrangements by creating a visual discourse that startles and disturbs. Through online depictions of themselves and their bodies, the twenty-something women often defied feminine and "girlie" traditions and displayed themselves as capable and assertive with Web pages that underscored their expertise. In claiming this cyborg territory as their own, these women on the Web—Laurel Gilbert and Crystal Kile (1996) tell us that "grrrls have attitude, girls don't"—clothed themselves in "attitude" and, as Haraway (1991) aptly states, committed their cyborg selves to "partiality, irony, intimacy, and perversity" (151). Of course, we continue to remind ourselves that these women were attending technologically privileged U.S. institutions of higher education and had both the credentials and the resources to build and maintain sophisticated Web spaces.

International, and Not So International, Feminist Web Sites

In the remainder of this chapter, we turn our gaze to women in other countries and explore how they represent themselves visually and verbally on the Web. We attempt, in part, to answer the question we posed early in this chapter: Or is it? That is, do the U.S. Web pages we examined display the complex geography of identity to which Friedman (1998) alludes as demonstrating multiple subject positions suffused with the cultural practices of "ethnicity, class, sexuality, religion and national origin, et cetera and so forth" (21)? Seemingly off the cuff, this question marks our uneasiness about the observations we made in our first two discussions of women's online lives—marks our sense that certain critical observations were not open to us because of our own im-

| **Figure 10.1** |

Russian WebGirls.

mersion in the Web scene. Thus, we now turn to international feminist Web sites for interrogation and contrast. Such a contrast should operate, as Friedman argues, as "a kind of categorical 'travel' that denaturalizes 'home,' bringing into visibility many of the cultural constructions we take as 'natural'" (114). As we proceed, we see boundary crossings and blurred identities as these women's sites relate to nation-states, geographical locations, cultures, and ethnicities. Enabled by technology, the women occupy subject positions that cross national and ethnic boundaries. Identity becomes a cyberhybridity. The identities that these women, no longer fully defined by history or geography, carve out for themselves are multiple, at once Russian or European but participating too in the marketplace economy of the Web dominated by Americans.

To illustrate this concept of cyberhybridity, we turn to a Russian Web site created and maintained by a group of women who hail primarily from St. Petersburg (see figure 10.1). Sporting the glitzy title of Russian Web Girls (2000), the site seems to function as an e-zine and meeting place for mostly Russian women and men. Its express purpose is to "break down the stereotype of Russian women." The site offers a choice of languages. The bold use of color—solid communist red, with a sidebar blaring "Russian Women Unite"—at once asserts a Bolshevik and proletarian identity before immediately undercutting it with a text that reads: "When Chanel redesigns the classic suit, the course of fashion changes for the next decade. And when five women get

| **Figure 10.2** |

Russian WebGirl Gallery.

together to break down the stereotype of Russian women, the world will never look at them again the same way. Russian WebGirls was born to make way for a new view of the Russian Woman—professional, beautiful, smart, sexy, multi-talented." Identifying and illustrating common stereotypes of Russian women, the Russian WebGirls then play with the images, asking viewers to choose from Babushka Gallery "which image better reflects a Russian woman": the renditions crafted by Art Lebedev, Inge Grape, Lena Secrest, Dasha Ziborova, or Mel DiaGiacomo (see figure 10.2)? Mel DiaGiacomo? A little bit of an Italian nationality enters the scene here too. Each image, when clicked upon, is accompanied by an animated babushka'ed figure who walks energetically alongside the chosen representation. Here, for example, with muscled women, the babushka'ed figure trots along the beach (figure 10.3) and then beside one of the nested Russian dolls, who just happens to be smoking Marlboro cigarettes (figure 10.4). In a later issue at the site, it's announced that Art Lebedev is the winner of the contest at the Russian WebGirls Gallery. The women announce that "[a]ccording to your votes, he had the best representation of a Russian Woman." He? He. Viewers do not have to be women to enter contests or participate in any other way at this feminist site.

Also in these pages are other babushka'ed figures who tell a wonderful story, which begins:

| **Figure 10.3** |

Babushka on Beach.

| **Figure 10.4** |

Babushka and Marlboros.

Having safely landed on the American land, babushkas unfastened their wings, hid them into their pouches/knapsacks, kotomki, and headed, over the Brooklyn Bridge, straight to the isle of Manhattan. The picture that opened to their eyes was like nothing they imagined back in a far-away village of Vaskovo. All around them

babushkas saw such a bewildering
abundance—of wonderful and exotic goods,
of strange and delicious foods—that looking
at all this, they felt a terrible urge to try
everything, taste everything, and bring as
much as would fit into their old kotomki
back home to their grandchildren.

Easy to say, harder to do, because our
babushkas had very little money, in fact,
none whatsoever. This started them thinking
about some serious money making. They
thought hard, for a day, two, and finally, on
the third day remembered about a traditional
Vaskovo art—the wings' weaving.

The creators of the site go on to relate that they asked their relatives in Vaskova to sell the cow and piglet and send them money to set up a business, "Fly-Away-Incorporated," specializing in wings manufacturing. Their advertising pitch reads:

Our Company "Fly-A-Way, Inc." Welcomes You!
We know the secret of FLYING. Our WINGS
will allow you to discover the new horizons,
to reach the unreachable heights
as well as
to burn 500 calories per hour by wing-waving.
WAVE AN OUNCE—LOSE TEN POUNDS
WINGS are BEAUTIFUL, WINGS are HIP,
WINGS will take you HIGH.

The business was a tremendous success, and the babushkas weaved all day and night, not unlike Haraway's oppositional cyborgs, while singing a song learned from their own babushkas in Vaskovo. Soon, they tell us, "Fly-A-Way, Inc." was so successful that it moved to the World Trade Center. The story, to be continued, momentarily stops as the author of the piece, Dasha Ziborova, writes: "With all their innovative ideas, babushkas became rich and famous. They were interviewed, photographed, invited to the talk-shows."

This site crosses national and even feminist boundaries in not so predictable ways. An Italian, Mel DiaGiacomo, a *he*, Art Lebedev, who created the feminist or feminine image that most appealed to voters at the site, and the wonderfully kooky Babushka story juxtaposing the Brooklyn Bridge and the Russian village of Vaskova and combining images of the Russian pig Petruska with images of wings that allow women to burn 500 calories an hour and to end up finally on talk shows, the epitome of success—all compete for viewers' attention. But the zany and playful identities we encounter here are only part of the narrative. The Russian WebGirls site, for example, was not created in Russia. Instead it owes its origins to six mostly Russian women, all of whom live in the United States, primarily New York.

All but one are fairly recent arrivals in the United States, and are, to our eyes, amazingly talented. Among them is a woman who owned an Internet service provider in Russia and continues it in New York; three are artists; one an engineer; another a jeweler, portrait artist, industrial designer, and cartoonist. The woman who came to the United States at a young age is the English translator, a journalist at the *Riverdale Press* in the Bronx who has aspirations to go to Russia as a foreign correspondent. There is also a Russian translator by the name of Inna Kolobova, who, unsurprisingly, dwells not in Russia but in Dallas, Texas.

All in all, then, the women at this site take traditional representations of Russian women and articulate them anew. The building of identity becomes a project of establishing new lives in which they search for the "we" and "us" that give them meaning. Cut off from their historical context, the women nevertheless playfully adopt and adapt folkloric representations of Russian life and place them side by side with parodic representations of the American or westernized woman's dream (lose weight, get rich, appear on talk shows) shaped by their own ambitions. For them, resistance is not so much finding new subject positions from which to speak as in recognizing the power structures in which they already participate and enacting them in different ways. Following Friedman (1998), we would argue that their cyberhybridity, "as a discourse of identity . . . depends materially, as well as figuratively, on movement through space, from one part of the globe to another" (24). It also, however, brings together the locality of place—New York, Russia, Vaskova—with a cyberspace inhabited by active participants from Russia, the United States, and other parts of the globe.

A second feminist site originates in Aachen, Germany, and, like the Russian WebGirls site, is intended as a women's resource site but, in this case, a resource for professional, working women or "woman entrepreneurs." WyberNetz (2000) calls itself the *europäschen Treffpunkt für Frauen*, or European meeting place for women. Its owner, Ann-Bettina Schmitz, and ABS Web-Publishing, tells us:

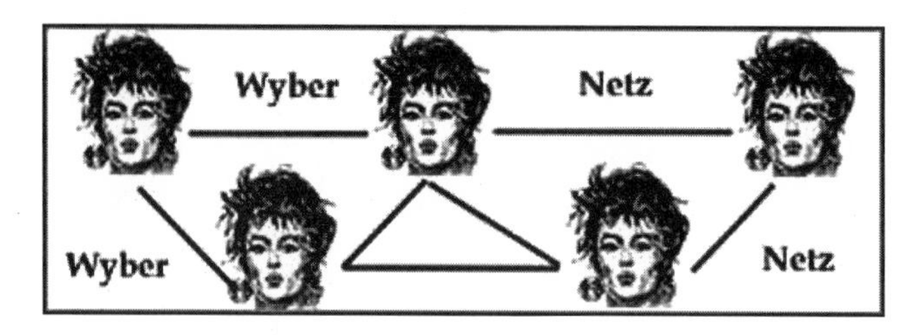

| **Figure 10.5** |

WyberNetz.

We are a German based woman-owned full web service provider. We have set up shop in Germany 4/1/96 and since then have been growing steadily. Maybe you've already heard of or even visited one of our international award winning sites WyberNetz and Showcase?
As a woman business owner you shouldn't miss listing your company—for free of course—at Women Entrepreneurs at WyberNetz . ABS Web-Publishing is located at Aachen, a nice old town in the west of Germany near the border to the Netherlands and Belgium. We have specialized in international networking and promotion of businesses. As we are a woman-owned company we take a special interest in the promotion of other women in business.

WyberNetz, like the Russian site, gives us a choice of languages (in this case German or English instead of Russian or English). And note that the name itself, WyberNetz, is a play on the German *Weib*, loosely translated into English as "wench," but then crosses over to the English (although Greek-derived) *cyber*, creating the sound-alike *Wyber* and a go-for-it assertive attitude to characterize the professional women participating in the space of WyberNetz.

When viewers go back to the home page of WyberNetz and click on the Women Entrepreneur link, they find a request for women entrepreneurs to submit an advertisement for their own business, and although the site is ostensibly a "European meeting place," women entrepreneurs from the United States have already established themselves here. There is an advertisement for "Little Prince and Princess Online Store" (2002) a business owned totally by women, we're told, and located in Tampa, Florida. Pastel colors of yellow, lavender, pink, and blue blend together invitingly, celebrating motherhood and tempting viewers to buy—by European standards—reasonably priced American baby clothes, probably sewn by women working in parts of Asia.

Another WyberNetz link, private homepages, invites viewers into the lives of many interesting and talented women from around the world but more likely than not from one of the Western industrialized countries. Among them is Soraida Martinez. Every-

| **Figure 10.6** |
Soraida.

thing at her site (Martinez 2002) indicates that she's a very successful artist and businesswoman: She gives exhibitions, visits children in school, and holds a seat on the New Jersey Arts Council. A Latina painter of Puerto Rican heritage, born in Harlem, New York, Soraida now lives in New Jersey. Looking straight out at us from her picture, she tells us that she is looking for someone to help her publish her Verdadism art book. "Verdadism," according to Soraida, is a term she invented from the Spanish word for "truth" and the English suffix for "theory," but there is more. In Verdadism, Soraida announces, "paintings are juxtaposed with written social commentaries in an effort to precipitate social change." With these words, then, Soraida Martinez proclaims herself an activist for social justice and shapes herself on the Web as an open advocate for social change. Adopting what we might call a cyborgian stance, Soraida becomes a storyteller through her art. Haraway (1991) notes that "[c]yborg writing is about the power to survive, not on the basis of original innocence, but on the basis of seizing the tools to mark the world that marked them as other" (175). Soraida takes up the artist's brush and the activist's stance and writes herself onto the World Wide Web. Here we begin to see how Web pages can be used for open activism even if that activism is tied securely to the more commercial occupation of selling a book of art. In fact, Web pages seem to be fertile spaces for bringing together the crass commercialism of Western society with subtle and not so subtle efforts to transform current configurations of society.

This sampling of current international and not so international feminist sites "manned" by talented, "global" women demands our attention all the more for the contradictions and competing discourses that mark the sites and that finally enable women to stage particular cultural and social identities. The first, is created by several Russian women, most of whom are recent emigrés to New York, who reinforce and dispel simultaneously common stereotypes of Russian women and who tell us wonderfully quirky counternarratives in the process. The other is a Web site for professional women run out of Aachen, Germany, but whose entrepreneurs sell baby clothes from Tampa, Florida, or exhibit and sell Verdadism Art, from Camden, New Jersey, in an overt effort to change American social institutions through artistic endeavors.

What are we to make of these stories that sometimes give us gendered representations but that also add women's feminist voices to the global conversation? And how do they add to or contradict insights from our previous studies of women online?

We might say that these sites serve primarily to feed colonial narratives with a corporate presence ubiquitous and firmly entrenched in Web pages and that they ostensibly multiply American markets while at the same time representing themselves as international or global women's Web sites. Cathryn McConaghy and Ilana Snyder (2000) argue, for example, that "similarities in architecture, design, use of color, relationship of text to image, the content, the political and cultural affiliations of web sites" suggest that the "commodification of the Web is a well rehearsed American, if not more broadly Western, habit" (85). We might so conclude, but we think we'd be wrong. Granted that Americans have supplied the technology and the advertisements in some cases, we would argue nevertheless that the women at the sites are using the technologies for their own purposes: to reach out to other women and to make names for themselves as the Russian women do, or to promote woman entrepreneurs as the WyberNetz *Frauen* do. And, in doing so, like the American college students of our earlier study, they are constituting a cyborg identity enabled by technological privilege and their own considerable expertise in Web authoring practices.

Cyberhybridities at Home and Abroad

Echoing Haraway, Anne Balsamo (1996) tells us that "[c]yborg bodies are definitionally transgressive of a dominant culture order, not so much because of their 'constructed' nature, but rather because of their hybrid design" (11). In this case, she's referring to cyborgs as part human, part machine, but the cyborg women represented in the Web pages discussed in this chapter transgress national boundaries to achieve multiple goals, constructing cyborgian, hybrid identities in the process. Homi Bhabha (1994) has suggested that "the display of hybridity—its peculiar 'replication'—terrorizes authority with the ruse of recognition, its mimicry, its mockery" (115). In cyborg fashion, the Russian WebGirls mock and mimic American and Russian cultural narratives, all the while constucting oppositional identities, bringing together women and computing culture, selling their wares, and building electronic social networks. Although there's no question that these Web spaces are for the privileged international set regardless of ethnic origins, we would like to view these spaces as a "third space," to use Homi Bhabha's term: an in-between space that "entertains difference without an assumed or imposed hierarchy" (4). Bhabha writes that "[t]hese in-between spaces provide the terrain for elaborating strategies of selfhood—singular or communal—that initiate new signs of identity and innovative sites of collaboration, and contestation, in the act of defining the idea of society itself" (1–2). These women, we would argue, are using these in-between spaces—these electronic third spaces—as locations in which to articulate meaning and identity through online writing practices, as spaces where meaning is, moreover, creatively constructed, rather than simply found. In their intermingling of various cultures, moreover, and in their play with the locale of place, they celebrate cyberhybridity as a powerful force for change.

With these women's web sites, then, we begin to see another possible trajectory of the global online story, one that relates to how women of multiple ethnic and national origins use (or not) the Web for their own goals and purposes. When we turn to these images, these online identities, and ask from what, by whom, and for what are they constructed, the answers are many and varied. Certainly ethnic and national considerations come into play: The folktales the Russian women tell on themselves in the Russian WebGirls' pages; the successful, Bridget Bardot pouting, professional head shot of a German woman entrepreneur featured in the logo of the WyberNetz site; and the Verdadism art that the Puerto Rican American Soraida Martinez creates. All of the women seem very aware of the ways in which others view their gender and ethnic origins, and all bend and shape these views to construct more complicated identities.

These women, like their American counterparts, write themselves visually onto the Web in multiple ways. Their use of advanced visuals, animation, and rich sets of resources all demand that viewers pay attention. The Russian WebGirls' efforts to create a new professional but sexy view of Russian women mirror the same characteristics that mark the young American websters of our earlier study. "Sexy" is part of this international feminist narrative; "dumb" isn't. And although Soraida is one of the few websters featured here who adds an explicit political explanation framing her art as a response to an American society that undervalues the artistic accomplishments of Puerto Ricans, African Americans, and women, we would argue that all the women represented are using Web technology to resist current cultural and social formations. Friedman (1998) has written that "migration through space materializes a movement through different cultures that effectively constitutes identity as the product of cultural grafting" (24). Although the migration reflected in the women's Web sites discussed in this chapter is enacted through cyberspace, it begins to constitute an identity that is more than Russian, American, European, German, or even Puerto Rican American. Thus, in adding their words and images to the Web, these women, like those in our earlier studies, have begun to weave an international feminist cyberquilt that can be a transformative force in the global settings in which we work and live. We see before us, then, an extensive, electronically networked landscape that—although a complicated and contested site for feminist Web practices—points to alternative opportunities for women in this new millennium.

Works Cited

Addison, Joanne, and Susan Hilligoss. 1999. "Technological Fronts: Lesbian Lives 'On the Line.'" In *Feminist Cyberscapes: Mapping Gendered Academic Spaces*, ed. Kristine Blair and Pamela Takayoshi, 21–40. Stamford, CT: Ablex.

Balsamo, Anne. 1996. *Technologies of the Gendered Body: Reading Cyborg Women.* Durham: Duke.

Bhabha, Homi K. 1994. *The Location of Culture.* New York: Routledge.

Escobar, Arturo. 1999. "Gender, Place and Networks: A Political Ecology of Cyberculture." In *Women@Internet: Creating New Cultures in Cyberspace*, ed. Wendy Harcourt, 31–54. New York: Zed Books.

Friedman, Susan Stanford. 1998. *Mappings: Feminism and the Cultural Geographies of Encounter.* Princeton: Princeton University Press.

Gilbert, Laurel, and Crystal Kile. 1996. *Surfer Grrls: Look, Ethel! An Internet Guide for Us.* Seattle: Seal Press.

Haraway, Donna J. 1991. *Simians, Cyborgs, and Women: The Reinvention of Nature.* New York: Routledge.

Hawisher, Gail E., and Patricia Sullivan. 1998. "Women on the Networks: Searching for E-Spaces of Their Own." In *In Other Words: Feminism and Composition*, ed. Susan Jarratt and Lynn Worsham, 172–197. New York: Modern Languages Association.

Hawisher, Gail E., and Patricia Sullivan. 1999. "Fleeting Images: Women Visually Writing the Web." In *Passions, Pedagogies, and 21st Century Technologies*, ed. Gail E. Hawisher and Cynthia L. Selfe, 268–291. Logan: Utah State University Press.

Jessup, Emily. 1991. "Feminism and Computers in Composition Instruction." In *Evolving Perspectives on Computers and Composition Studies: Questions for the 1990s*, ed. Gail E. Hawisher and Cynthia L. Selfe, 336–355. Urbana, IL: National Council of Teachers of English.

Kress, Gunther. 1999. "'English' at the Crossroads: Rethinking Curricula of Communication in the Context of the Turn to the Visual." In *Passions, Pedagogies, and 21st Century Technologies*, ed. Gail E. Hawisher and Cynthia L. Selfe, 66–88. Logan: Utah State University Press.

"Little Prince and Princess Online Store." 2002. RoyalBaby Web Site. Available: <http://www.royalbaby.com> (accessed February 12, 2002).

Martinez, Soraida. 2002. Soraida's Web Site. Available: <http://www.soraida.com> (accessed February 12, 2002).

McConaghy, Cathryn, and Ilana Snyder. 2000. "Working the Web in Postcolonial Australia." In *Global Literacies and the World Wide Web*, ed. Gail E. Hawisher and Cynthia L. Selfe, 74–92. London: Routledge.

Purves, Alan C. 1998. *The Web of Text and the Web of God: An Essay on the Third Information Revolution.* New York: Guilford.

"Russian WebGirls." 2000. Russian WebGirls Web Site. Available: <http://www.russianwebgirls.com/> (accessed September 8, 2000).

Selfe, Cynthia L. 1992. "Complicating Our Vision of Technology and Writing: What We Have Learned in the First Ten Years." Keynote address presented at Eighth Computers and Writing Conference, Indiana University–Purdue University, Indianapolis, May 3.

Selfe, Cynthia L. 1999. "Lest We Think the Revolution is a Revolution: Images of Technology and the Nature of Change." In *Passions, Pedagogies, and 21st Century Technologies*, ed. Gail E. Hawisher and Cynthia L. Selfe, 292–322. Logan: Utah State University Press.

Sloane, Sarah J. 1999. "Postmodernist Looks at the Body Electric: E-mail, Female, and Hijra." In *Feminist Cyberscapes: Mapping Gendered Academic Spaces*, ed. Kristine Blair and Pamela Takayoshi, 41–61. Stamford, CT: Ablex.

Spender, Dale. 1995. *Nattering on the Net.* Melbourne, Australia: Spinifex.

Sullivan, Laura. 1997. "Cyberbabes: (Self-)Representation of Women and the Virtual Male Gaze." *Computers and Composition* 14, no. 2:189–204.

Taylor, H. Jeanie, Cheris Kramarae, and Maureen Ebben, Eds. 1993. *Women, Information Technology, and Scholarship.* Urbana, IL: Women, Information Technology, and Scholarship Colloquium.

"WyberNetz." 2002. WyberNetz Web site. Available: <http://www.web-publishing.com/WyberNetz/hello.htm> (accessed February 12, 2002).

11

UNHEIMLICH MANEUVER: SELF-IMAGE AND IDENTIFICATORY PRACTICE IN VIRTUAL REALITY ENVIRONMENTS

Alice Crawford

> Werk des Gesichts ist getan
> tut nun Herz-Werk
> an den Bildern in dir, jenen gefangenen; denn du
> überwältigtest sie: aber nun kennst du si nicht.[1]
> —Rainer Maria Rilke, "Wendung"

> We are beings who are looked at, in the spectacle of the world.
> —Jacques Lacan, *The Four Fundamental Concepts of Psychoanalysis*

As Duke Nukem, Alien Pig-Cop killer and anti-porn crusader, you cruise into the men's room between rounds and contemplate your well-muscled blonde visage in the mirror. Your image gazes lovingly back at you and purrs: "Damn, I look good!" Then, because you are playing the popular first-person-shooter video game, rather than living your own, more pacific life, it's back to the labyrinth to blow the knees off of more white slavers from outer space. This moment in *Duke Nukem,* when you are hailed from the screen by what is, in essence, your ideal-ego, which is also the first-person character you've been "inhabiting" through the joystick and, if you're like most video game players, through sitting very close to a rather large television screen, is *unheimlich* in the extreme if one is prone to such sensations. For most players, I imagine, the moment is rather banal. This is, perhaps, just as uncanny, for as we become more accustomed to inhabiting points of view in increasingly immersive simulated environments, it may become commonplace to experience temporary but compelling "self"-identification with characters whose visual characteristics[2] are far removed from one's real-world, embodied attributes.

I begin with an anecdote about this moment in *Duke Nukem* because it carries within it, if we care to tease them out, long strands of questions now under debate in both popular and academic discussions of emergent image-making technologies, most pointedly, questions regarding the nature of the relationship between self-identity and media images. If it is possible to experience a *frisson* of self-(mis)recognition while playing a video game, how far might self-identity be morphed when one is inhabiting a character in a more immersive environment like that of so-called virtual reality (VR)? What would this mean, in terms of relations with ourselves and others?

VR: One Caveat and Three Strains of Speculation

Although the term "virtual reality" has long ago begun to carry with it the whiff of a dead fad, it still serves to pick out a complex (and continually shifting) constellation of technologies and practices actually under development: at its most basic, VR can be described as an array of devices[3] that together form a human-computer interface through which the "spectator" (or better yet, "participant") is immersed in electronically simulated sensory inputs. This immersion works to bring about the sensation of a first-person point of view within an enveloping, artificial environment that can be navigated and manipulated. Howard Rheingold (1999), a popular commentator on the VR phenomenon, describes the VR interface in a refreshingly simple fashion: "Virtual reality is . . . a simulator, but instead of looking at a flat, two-dimensional screen and operating a joystick, the person who experiences VR is surrounded by a three-dimensional computer-generated representation, and is able to move around in the virtual world and see it from different angles, to reach into it, grab it, and reshape it" (17). It is this sort of immersive human-computer interface that will be passing under the label of "VR" here, and it is this sort of interface that offers up such an interesting puzzle in terms of understanding the organization of the human subject in the context of the field of vision, for here the structures of the visual environment in which we recognize ourselves and others are undergoing a rather remarkable reformulation.

Even a cursory glance around popular cultural and the general run of critical responses to the possibilities of identity formation in VR might well give a person the impression that this is largely a disastrous development. Aside from a few pockets of enthusiasm for endless shape shifting, mostly found in sci-fi-ish "lifestyle" magazines such as *Mondo 2000* and *21-C* or in the pages of high postmodernist theory, anxiety is more the norm when it comes to speculations regarding the more powerful, perhaps even intoxicating forms of identification: forms that would be enabled through interaction with increasingly immersive, multisensory interfaces. In response to this anxiety, many critics of the emerging media have joined in a long-standing tradition within media scholarship (inherited from studies of classical narrative cinema, in large part, in which the question of identity/identification looms large). These critics take a decidedly ascetic stance toward the visual images on offer in popular media and prescribe a defensive strategy of rational, dispassionate distanciation from the popular visual.[4] It is suggested that this strategy will, if followed with enough self-discipline and rigor, draw us out of the murky twilight of the visual register and help cure the ills of the body politic.

Although anxiety regarding the role of visuality in constituting the subject has circulated around the technologies of photography, film, and television, the feeling of unease is even more acute for emergent technologies of vision. This is not surprising, really, given that the nature of the encounter with the visual field is notably different in VR than with previous technologically enabled forms of visuality: unprecedented and strange, even *unheimlich*, as I have suggested. In "Supernatural Future," Sean Cubitt (1996) expresses a fairly common understanding of what such an encounter might look like:

Indeed, the HCI [human-computer interface] seems to be configured around an intensification of the psychic relations formed as identification in cinema, where the dialectic of public and private, of social ritual and intimate fantasy, powers the formation of glamour, returned to the private sphere in the foreshortened space of the domestic television, and now moving to the hunched, one-on-one foetal curl of the body engrossed in the VDU [video display unit], with the promise of ever intensely individuated interfaces in VR. This hyper-individuation works by a similar regression to that evoked by cinema, where the dialectic of identification regresses its audience to the moment of ego formation. In the HCI that regression is carried even further back. (248)

Cubitt is not alone in suggesting that an immersive medium such as VR will not only enable but provoke forms of identification in which the "glamour" (in the oldest sense of the word) of images on screen will work a bewitching spell on our psyches. This leads to Cubitt's scenario of primary narcissism recovered in adulthood, or, as is more commonly proposed, to overidentification with the idealized images of the dominant and dominating culture of standardized media representations.

Such representations include, for example, images like Aryan, hard-bodied Duke Nukem, identification with which would, it is argued, at best strongly interpellate the spectator/participant into bad subject positions, and at worst provoke a psychotic misrecognition of the self as an all-powerful, protean ideal. In pop cultural meditations on VR, examples of such psychotic episodes are a common trope. Consider *Lawnmower Man,* an early and much-cited movie about VR in which a simple-minded handyman is transformed into a powerful VR daemon. Intoxicated with his new powers, he attempts to destroy the world and to take women by force and engages in various other demented endeavors. The movie's tag line, which appeared on the posters, was "God made him simple: Science made him a God." But of course, he wasn't really a God in the end. He just went insane.[5] Even Mark Pesce (1994), the author of VRML (virtual reality modeling language), the code often used to build 3-D Web worlds, has described what seems to be an almost built-in capacity to provoke psychotic moments as the "pathogenic ontology" of VR and warns that VR should be used with caution (29).

Speculation about VR and identity is not, of course, limited to such dire prognostication. There is also a sizable contingent of popular and not so popular critics who are eager (perhaps *all too* eager sometimes) to see the boundaries of the self break down into fluid, polymorphous algorithms in which the mind somehow escapes the "meat" of the body to sail off into the "smooth space" of the matrix. Not surprisingly, some of the most boosterish analysis comes from people working in the fledgling industry of VR and in the margins of the technology press. Take, for example, Jaron Lanier, who has become something of a poster boy for new improved VR identity. Lanier, who worked at the Atari labs during their glory days and was instrumental in developing some proto-VR technologies,[6] can often be found lamenting the fact that our identities are bound by the mundane constraints of the physical world: "We are actually extremely limited. The earlier back into my childhood I remember, the more I remember an internal feeling of an infinite possibility for sensation and perception and form and the frustration of reconciling this with the physical world outside which was very, very fixed, very dull, and very frustrating—really something like a prison" (Lanier, 1999, 242). Lanier and his cohort have suggested that, rather than attempting to establish our identities in the embodied world, we should turn our gaze to the screen, where we can manipulate our self-images endlessly in a virtual world where anything is possible (and nothing is forbidden). When it comes to understanding the way in which subjectivity and the world of visual representation operates, this approach is a through-the-looking-glass twin of the ascetic, even iconoclastic point of view outlined above. On this side of the looking glass, the kind of regressive, narcissistic relationship to the technologies of image making that Cubitt et al. abjure has become a lifestyle objective.

At the same time, a more measured and productive strain of less technophobic literature on the subject is being written by feminist scholars[7] following in the footsteps of Donna Haraway. Haraway, in her practically omnipresent "Manifesto for Cyborgs" (1985), suggested that the blurring of boundaries between human and machine, interior and exterior, self and other might in fact be a change for the better. Human, embodied identity, so the story goes, has too often been the site of gendered, racial, and class oppression. Leave it behind, and more egalitarian prospects open up. Although I am somewhat sympathetic to the appeals of "becoming cyborg" that this tradition of scholarship tends to promote (at least as a theoretical position, if not an actual lifestyle), I am wary of the suggestion that the process of identification needs to break free from its moorings in human embodiment to serve a progressive political agenda. When N. Katherine Hayles (1998) suggests that, in VR, "subjectivity is dispersed throughout the cybernetic circuit . . . the boundaries of self are defined less by the skin than by the feedback loops connecting body and simulation in a techno-bio-integrated circuit" (81), she seems rather sanguine about the prospect.[8] I, on the other hand, cannot help but wonder whether it would not be more productive to keep our focus more firmly on ways in which VR technologies might be used to enhance our embodied intersubjective relations and our relations with our own embodied selves.

To summarize briefly and to simplify a bit, speculation regarding the formation of identity in VR environments tends to be characterized by one of the following assumptions regarding the formation of identity in immersive media environment:

1. Our encounters with VR interface will lead us into a narcissistic regression with psychotic episodes, and this will be bad.
2. Same as 1, only this will be good.
3. VR immersion will disconnect the identificatory process from embodiment, producing mobile, fluid subjectivities that play themselves out primarily "on screen"—in the virtual locales of the interface, rather than in our lives here in the off screen world of fleshly creatures.

On this last perspective narcissism is displaced as central trope, giving way to an understanding of VR as a place in which fluid ego boundaries become an almost foregone conclusion. Although this final perspective generally avoids a simplistic understanding of identificatory processes, it tends to focus on identification as it is played out in a supposed transition to a posthuman form of subjectivity rather than focusing on how emergent technologies might instead serve to make us more fully human in our lives off screen.

———

Where Do We Go from Here?

Given how differently each of these scenarios would play out in the socius at large and given all that is at stake, it seems wise to keep in mind the radical futurity of VR, the fact that as a technology and, especially, as a set of cultural practices, VR is under invention at this moment in time and that the manner in which we imagine its uses will be determinative[9] of what it will be. Clearly, we cannot simply look at the arrays of devices and implements, chips and wires, currently in circulation and try to divine what they will do, as if we were reading the future off of goat entrails.[10] As Carolyn Marvin (1988) has convincingly argued, "media are not fixed natural objects; they have no natural edges. They are constructed complexes of habits, beliefs, and procedures embedded in elaborate cultural codes of communication" (4). Constant reminders that media are "cultural technologies"[11] are needed to fend off the surging tide of technologically determinist accounts that purport to tell us what we "will" be doing in the future with these contraptions. All of this aside, my argument here is rather simple: that it is not only possible, but likely, that there will be some notable changes in the relationship between the field of vision as it is articulated by emergent image-making technologies and the way the psychic process of identification takes place within this field. Furthermore, it is our responsibility at this juncture to imagine tactics for shaping these changes in identificatory practice so that they may serve to liberate us somewhat from the tyranny of dominant modes of idealization in our lives, both on the screen and, most importantly, *away from it.* My suggestions below seek to address some of the drawbacks of the perspectives I have outlined above, particularly the first, "iconoclastic" position, which offers only suspicion regarding the potentialities of visuality and identification in VR environments. The second, "happy narcissist" approach might seem a bit too callow to take very seriously, really, but it does have a fair amount of currency. It also clearly perpetuates a similarly narrow and solipsistic notion of what identification is. The third perspective, that which dreams of the "cyborg-self," tends to focus on the way in which VR encounters might be productive of a cyborg-self that forms its identity not so much in relation to images of the human, but to the machinic and cybernetic. My remarks in the remainder of the chapter will offer a model for understanding subjectivity and visuality that attempts to avoid these drawbacks. I make my suggestions as a way of entertaining the possibility that, under the right circumstances, visual practices in the medium of VR might enable new ways of *seeing* and *being seen* that could have a salubrious impact on our lives both on and off the screen, in our not-yet-and-hopefully-never-post-human flesh. A reconsideration of what is entailed in the process of identification will be productive in amending some oversights in the perspectives outlined above.

The Subject of Vision

Although it has become rather commonplace in a number of disciplines to claim that subjectivity is formed in relation to language and textual practices of representation, less attention has been paid to the mechanisms though which subjectivity arrays itself with respect to the world of images. What scant attention has been given has, historically, been dismissive to the point that W. J. T. Mitchell (1991) has characterized the bulk of speculations regarding the relationship of the visual world to subject formation as follows:

> Is this subject primarily constituted by language or by imaging? by invisible, spiritual, inward signs, or by visible, tangible, outward gestures? The traditional answer to these very traditional questions has been to privilege the linguistic: man is the speaking animal. The image is the medium of the subhuman, the savage, the "dumb" animal, the child, the woman. (323)

In puzzling out the workings of visuality and identity, it has been difficult to escape the influence of Jacques Lacan, given the centrality of the question of vision in his writings and lectures. Though I am sympathetic to Laura Kipnis's (1993) observation that the feminist romance with Lacan "strongly suggests the Harlequin formula: the hero may be, on the surface, rude, sexist, and self-absorbed, but it is he alone who knows the truth of the heroine's desire" (103), it is also the case that Lacanian categories of analysis offer a rich and resonant vocabulary with which to discuss the formation of subjectivity in the field of vision, or what Lacan has termed the "Imaginary."

In Lacan we find a treatment of the organization of the human subject in which vision plays a key role. Most famously, Lacan's (1977) account of the "mirror stage" posits the recognition (or misrecognition) of the image as a reflection of the self as a founding moment in the development of the ego. This moment takes place, so the story goes, in the infant's encounter with the mirror image of herself, in which she appears as a self-contained and integrated whole, a moment of (mis) recognition that "situates the ego, before its social determination, in a fictional direction, which will always remain irreducible for the individual alone" (2). A coherent sense of selfhood, on this account, has its origin in the apprehension of an image of a human body: the child's own. That the illusion of wholeness and self-sufficiency is just that, an illusion, does not change the primacy of visual relations in this account of the formation of the self.

Furthermore, what resonates here is what Lacan refers to as the "irreducibility" of self-identification with the image of the body; in the case of the mirror stage narrative, this image is the indexically motivated sign of the child's own embodied psyche, "indexical" here referring to the fact that the connection between the visual signifier (the image in the mirror) and the signified (the child's body) is not arbitrary. The image is related to the subject in this instance as a material trace, in an odd but tangible sense, of her own, unique fleshly body, a body that moves as she moves and generally responds in a congruent manner with the tactile and kinesthetic sensations that also tell her where she is, and where her boundaries are, etc. I want to emphasize this moment of "irreducibility," which is often overlooked in accounts of identity formation that propose the existence of a fully mobile subject. The connection that is drawn between the mirror image, bodily sensation, and a sense of self, rather than being yet another arbitrary relation open to rearticulation at any moment, is grounded in a synaesthetic moment made possible only through inhabiting a particular set of embodied coordinates. This is quite different from the relation of identity to language, in which the signifier "I," for example, applies as well to me as to you as to anyone else. In the image, there is something more powerfully founding of a stable (if fictive, in a sense) form of identification than the realm of language (Lacan's "symbolic") can offer.

Though it is reasonable to assume that a founding moment of subjectivity occurs in our encounter with a self-image,[12] it would be inappropriate to infer that the way in which the relation with images transpires in further encounters with visuality is uniform. In fact, it seems clear that after this initial encounter, identificatory processes proceed in two starkly different manners. The differences between these two forms of identification are key to understanding how the mechanisms of a self-identity founded in the visual can map quite differently onto intersubjective relations. Reconsidering the distinction between these modes will help in assessing the adequacy of the various accounts of visual identification in VR outlined above.

Narcissistic Identification and the Normative Ego

Identification can be carried out in two radically different modes or, more accurately, *is* carried out in radically different modes at different moments by all of us. The first alternative is graspingly narcissistic, a form of identification that has also been termed "incorporative" or "idiopathic," all of these terms emphasizing the literally self-centered, centripetal nature of this form of identification. In this form of identification

———

the subject attempts to ascribe to itself whatever is deemed desirable in the visual register. It is this narcissistic striving to incorporate all of ideality—"I am the world/I can contain all"—that Cubitt (1996) refers to with respect to VR as "the primary narcissism of 'Her Majesty the Baby'" (248), in which the subject's only relation to the visual world is that of an ongoing reflection of her own sweet self. The existence of other subjectivities is of little consequence, if not denied outright, and relations with others take on the form of object relations. This is the same form of identification that, in a psychotic mode, provokes the subject to misrecognize herself as an all-powerful, all-encompassing being. This is an identificatory process that would produce the monadic, self-absorbed shape shifters (and Lawnmower Men) that many critics fear VR will enable.

In this process the normative ego strives desperately to be whole through its insistence on an identification with the ideal and a denial of lack. In *The Threshold of the Visible World*, Kaja Silverman (1996) has mapped out an exhaustive account of the many permutations of identificatory practice in which she usefully describes this incorporative form identification as playing out a psychic insistence on "the principle of an integral self." This principle, she argues, is "tantamount to an inexorable insistence upon sameness" (92) by the ego and is a form of psychic maneuvering of which critics of identification vis-à-vis media images are right to be wary, given its implications for intersubjective relations. The incorporating, narcissistic self not only ascribes inordinate value to itself but maintains an attitude toward other subjectivities that is defensive and even destructive. Silverman describes this attitude as a "hostile or colonizing relation to the realm of the other. Confronted with difference, the ostensibly coherent bodily ego will either reject it as an unacceptable 'mirror,' or reconstitute it in digestible terms" (92).

What is incorporated in narcissistic identification includes only the ideal and is founded upon strong, even murderous abjection, of the nonidealized both in ourselves and within the socius. As Judith Butler (1993) has pointed out with reference to the symbolic realm, the process of identification is in large part a negative process, in which certain identifications are enabled and others foreclosed or disavowed. She rightly emphasizes that the abjection on which self-identification is founded is always a fundamentally social (or antisocial) affair:

> The abject designates . . . precisely those 'unlivable' and 'uninhabitable' zones of social life which are nevertheless densely populated by those who do not enjoy the status of the subject, but whose living under the sign of the 'unlivable' is required to circumscribe the domain of the subject. This zone of uninhabitability will constitute the defining limit of the

subject's domain; it will constitute that site of dreaded identification against which—and by virtue of which—the domain of the subject will circumscribe its own claim to autonomy and to life. (3)

On this account, the all-or-nothing drive for wholeness that characterizes narcissistic identification is a constitutive force in relations of domination and exclusion in the realm of intersubjective relations. It is also an invitation to an endless oscillation between self-adoration and self-hatred as the subject momentarily achieves successful identification with the ideal, then inevitably realizes that she cannot quite measure up. The fact that this process is embedded in standardized cultural tropes makes its relation to real, historical structures of inequity even more clear.

In the context of understanding the potential impact of VR on identificatory practices, it is crucial to note that this form of identification operates by constantly parsing out (unconsciously) whom we recognize as subjects and does so along the lines of normative representational codes, the constantly reiterated hierarchies of images that circulate around us, increasingly in electronically mediated forms. All of this in mind, it is clearly wise to be a bit wary of the process of identification vis-à-vis the "popular visual," and it is this narcissistic model that is so often encountered in the more alarmist critiques of VR's potential effects on subject formation. To assume that all identification with idealized visual images takes place under the sign of narcissism, however, ignores the fact that this process is not always and only such a self-centered affair: Although identification with visual ideals opens the door to narcissism and *mis*recognition of the self as all-encompassing, it also opens the door to recognition of the other and makes possible intersubjective relations, which brings us to the second form of identification in question.

Heteropathic/Excorporative Identification

Opposing the narcissistic, incorporative psychic formation is a very different mode of identification variously termed "excorporative" or "heteropathic." In this mode the subject, rather than attempting to see all that is ideal in the visual register as an aspect of an all-consuming self, is capable of appreciating the ideal in others and, more importantly, recognizes the subjectivity of others who do not necessarily fit neatly within the parameters of normative codes of representation. It is with this second, more intersubjectively generous form of identification which Silverman is centrally concerned in her work on what she has called "the cinema of the productive look," a notion that, although not entirely well suited to describing a progressive approach to VR, has certain fundamental similarities to the attitude toward VR as a visual medium that I am suggesting. Silverman emphasizes the multiform nature of ideality as it appears in the process of identification to draw out the possibilities for seeing differently and ascribing value in ways that the normative ego would foreclose. Contrary to the conventional model of identification as a strictly narcissistic affair (the model media critics inherited from studies of classical narrative cinema), Silverman (1996) suggests a model of identification in which, through conscious, collective effort, the idealization that precipitates identification is located in *others*, rather than strictly in the self, and in the traditionally abjected, rather than solely along the lines of normative representational practice:

[I]deality is the single most powerful inducement for identification; we cannot idealize something without at the same time identifying with it. Idealization is therefore a crucial political tool, which can give us access to a whole range of new psychic relations. However, we cannot decide that we will henceforth idealize differently; that activity is primarily unconscious, and for the most part textually steered. We consequently need aesthetic works which will make it possible for us to idealize, and, so, to identify with bodies we would otherwise repudiate. (2)

What Silverman is pointing to here is an often-overlooked aspect of the process of identification: that an ideal remains a powerless abstraction until it has been psychically affirmed and that the affirmation of ideality is a profoundly intersubjective affair. Without the intersubjective relationship of looking and being looked back at and without being recognized by others as fitting or failing to fit within the parameters of idealized images, the endless procession of images produced by visual media mean nothing either in terms of self-identification or in the classification of others as ideal or abjected. As Silverman puts it, “we can only effect a satisfactory captation when we not only see ourselves, but *feel ourselves being seen* in the shape of a particular image” (57, emphasis added).

VR as a “Theater of the Productive Look”

Although Silverman’s “cinema of the productive look” is a strictly filmic affair, the intersubjective aspect of excorporative identification she elaborates is tremendously generative in understanding the liberatory potential of VR experiences. Unlike other visual media, such as photography, television, and film, VR is, at least potentially, a profoundly intersubjective affair. Brenda Laurel, an interface designer and critic, suggests understanding the dynamics of VR as more closely approximating theater[13] than anything else, given that its most interesting and compelling forms will be structured around interactions with other actors *within* the medium itself.

In Lacanian parlance, usually when it is suggested that captation occurs in the moment of "feeling ourselves being seen," the entity referred to as doing the looking is understood to be the strangely disembodied, almost animistic force known as "the gaze." Although the question of the gaze is a fascinating one, I believe we can safely leave it aside here, for it seems clear that captation also occurs with the kind of collective looking that takes place among human subjects. This kind of looking is not the same as the gaze, as a number of recent critics have rightly pointed out. I still argue, however, that the look, when performed collectively, can serve a similar function to the gaze. Namely, it can serve to affirm or deny the subject's self-identification with the ideal and can also, to a much greater extent than theories of the gaze often admit, affect what counts as ideal within the visual realm.

The notion of an *active* idealization should not be misunderstood as a sneaky importation of individual agency vis-à-vis the gaze. It is, on the contrary, a necessarily *collective* endeavor, attempting through the mediation of aesthetic texts and practices to shift the visual terms through which we apprehend the world. Idealization in *this* mode opens up identifications that would otherwise be foreclosed both by the standard representational practices of the dominant culture and by the imperatives of the normative ego. The gazelike function of the collective look has implications for VR as for no other medium: As an intersubjective medium, VR opens up intriguing possibilities for retooling the identificatory process because its immersive and intersubjective characteristics make it possible to experiment with "occupying a subject position that is antithetical to one's psychic formation/self" (Silverman, 1996, 91) in an environment in which one can be "captated" in a psychically powerful fashion, because one is *recognized by others* with whom one is interacting as occupying that subject position.

In this respect VR might potentially serve as a medium for collective experimentation with heteropathic forms of identification. This experimentation would have the capacity to be more affectively powerful than is possible in other electronic media. To illustrate my point here, I'm going to make a narrative leap to some rather famous remarks Lanier has made relating to this issue (though in a quite different context and toward a different end). Speculating about the use of VR as an instrument of self-transformation, Lanier emphasized the ability of VR to simulate "trading eyes" with another person, even swapping nervous and motor systems with another species, in Lanier's example, a lobster: "[t]he interesting thing about being a lobster is that you have extra limbs. . . . We found that by using bits of movement in the elbow and knee and factoring them together through a complex computer function, people easily learned to control those extra limbs. When we challenge our physical self-image, the nervous system responds very quickly" (quoted in Boddy 1994, 118). Lanier's remarks point out, in a somewhat exaggerated form perhaps, an often overlooked point in discussions of identificatory processes in emerging media, namely, that spectatorship in VR can potentially be quite different from spectatorship vis-à-vis film, television, or photography.

This possibility of inhabiting, temporarily, the visual and physical coordinates of the Other sets the experience apart. One not only imagines oneself in the image of the Other, as in film spectatorship, for example, but *is seen by others* in this guise. Within VR one might feel oneself being "photographed" by many looks as one occupies bodily coordinates radically other than those of one's embodied, off-screen life. The intersubjectivity of identification in this case would produce a more psychically resonant captation than occurs through other media. Cinema, television, and photography all offer images to idealize or abjure, but they do not incorporate real-time intersubjective relations of looking and being looked at in the way that the VR interface can do. In my thinking about this topic, I have been tempted to term these psychically powerful moments of captation-as-other as a form of "ludic psychosis," pointing to the destabilizing effects these moments might have on the normative ego. During captating moments of being recognized as *Other*, the necessity, naturalness, and inevitability of the particular position we normally occupy within the visual register would, I believe, be at least *opened* to a radical form of doubt.

Unlike the permanently deranging effects of ongoing psychosis, however, these moments of captation-as-other would be just that: moments. One always has to take the helmet off, jack out, or exit the teledildonic chamber eventually.[14] Even in the case of identification with idealized coordinates, the effects of "degoggling" could, under the right circumstances of collective practice around this medium, have a destabilizing effect on narcissistic forms of bodily identification. One minute you're the Duke, invincible and as perfect as only simulated blonde plastic can be, and then the next moment you're, well, you're yourself again. This wrenching removal from the scene of psychic captation in the ideal could, if grounded in an understanding of VR as a theatrical, intersubjective medium, clearly display one's irreducible distance from the array of images of the body that constitute dominant cultural ideals.

Under the right circumstances, such swapping of visual coordinates might allow us to rearticulate cultural standards of corporeal ideality productively by showing up the fictive and essentially arbitrary nature of representational norms and, hence, of identity in the imaginary register. As I see it, VR has the potential to bring into question the visual norms by which the gaze sorts us into desirable, undesirable, having and lacking, prized and disprized, and in an affectively powerful fashion. Accordingly, VR might allow us to play with difference in an unprecedented fashion, as it offers unique opportunities for producing visual texts that allow one to be displaced from oneself and facilitating the reading of visual representations "against the grain." In his typically boosterish style, Lanier once said that "VR is the ultimate lack of race or class distinctions, or any other form of pretense, since all form is variable" (quoted in Boddy 1994, 118). Now, clearly there is a certain naiveté to such a categorical (and voluntarist) statement. The variability of visual form in VR could, however, have some interesting effects, especially in the case of taking on the visual coordinates of culturally disprized "otherness," such as, to take an extreme example, those of the "lobster boy": Grady Stiles Jr., world-class sideshow freak. It would be something *quite* different—structurally, qualitatively, affectively—to watch a filmic representation of a culturally disprized body like the lobster boy's and to feel some identification with that subjective position than it would to be in an interactive environment in which you are seen by others as that body: "Look, it's Lobster Boy!" or, "Look, it's that homeless lady!" or "Look, it's Langston Hughes!"

Visually inhabiting a simulated subject position in an immersive, interactive environment is certainly not *identical* to the experience of inhabiting its off-line, embodied counterpart. Therefore, it would be a mistake to argue that to inhabit Langston Hughes's identity in such an environment would allow one to fully know what it would be like to live through that identity in all of its embodied complexity. Nevertheless, this partial inhabitation of the identity of another is a step in an interesting and useful direction, and one that VR could enable in an unprecedented fashion.

By providing an environment in which we can assume various prized and disprized identities in an affectively resonant fashion, VR might allow us to experience more profoundly the arbitrariness of individual location in the "language" of visual representations and the effects of the arbitrary cultural valorization of different bodily images and the abjection of others: male, female, black, white, queer, skinny, fat, poor, freak. It seems possible that these moments, perhaps momentarily, ludically psychotic in their deranging of normal codes of identification, could coax us to acknowledge the abyss that separates and always will separate self-identity from ideality, a relationship that Hal Foster (1996) has rather poignantly characterized as "the gap between imagined and actual body-images that yawns within each of us, the gap of (mis)recognition that we attempt to fill with fashion models and entertainment images every day and every night of our lives" (110). From this perspective, any encounter that would assist us in collectively recognizing culturally idealized representational norms as fictive, rather than natural or originary, would be sweet relief.

An Ethics of Spectatorship and Interface Design

Clearly, VR is not an *agape* machine: The technology itself won't inevitably produce destabilizing aesthetic experiences or bring about a radical reconsideration of our normative forms of identification and idealization in the visual register. What is required is an ethic of spectatorship that values such experiences and incorporates entertainment, escapism, and play in a collective search for the good life. (I think here of the ethos of the Cultural Front in the 1930s, a moment in which these elements came together for a brief time without immediately degenerating into arid, joyless pedagogy). On a practical level, there also needs to be a wide array of visual coordinates available for temporary occupation. Currently existing virtual environments, such as they are, are largely commercial spaces designed for shopping, or, at best, commercially created environments such as Habitat, World's Chat, or the wildly popular Sega Dreamcast system. Participants can select a predesigned "avatar" to represent them visually to others but cannot in any meaningful way map out their own visual coordinates and are largely offered innocuous and/or idealized avatars, Kyoko Date–like[15] cutie pies, muscular, lantern-jawed fellows with executive hair, pink bunnies, and the like.

To actualize VR's utopian possibilities, it is important not only that we collectively guide our use of the medium in order to play with something beyond banal ideality: Everyday citizens in the "bitsphere" (as the emerging "public" space of immersive media environments has been described) must also be technically and procedurally capable of downloading their own visual material versus having it dished out entirely by Lucasfilm or Disney and their cohort. As Rheingold (1991) has pointed out, "[a]t the heart of VR is an experience . . . and the problems inherent in creating artificial experiences are older than computers. While MIT and the Defense Department might know a thing or two about spurring new computer technologies, the center of the illusion industry is closer to Hollywood" (46). The design of virtual environments along standard Hollywood lines would, perhaps, offer occasional moments of *unheimlich* (mis)recognition but would clearly not be ideal for forming a contestatory cultural practice within the medium. Fortunately, this is a problem currently being addressed by a number of leading designers of VR environments, including Laurel (1992), who points out that, for virtual environments to have any meaningful variety of subject positions, the authoring tools for these environments probably need to be distributed to the users. The strategy, in her view, is "to make the authors of the content also the authors of the structure and the interface" (86), so that the builders of these environments aren't simply appropriating content and putting it into, as she puts it, "White Western" form and structure.

It may be impossible, in the end, to create authoring tools entirely free from cultural bias, but it's a productive ideal to drive toward, even if only asymptotically approachable. If it *does* in fact become possible for anyone with access to a computer[16] to add her visual information to the gallery of avatars on offer, this will not only reorganize the positions of producer and consumer of media images, participant and spectator, but will also allow for visual practices that may better serve as a heuristic for progressive political practice.

Playing with the *Unheimlich:* Ludic Psychosis

In sum, much of the critical work on VR oversimplifies the psychic processes involved in visual identification by emphasizing only the narcissistic and psychotic aspects of a phenomenon that is in fact quite complex and multiform. There has also been little attention given to crucial differences between how VR might inflect the affective experience of images in comparison to other visual media, such as film, television, or photography. With these differences in mind, it seems clear that identification in VR cannot adequately be described by theories of spectatorship developed for earlier media forms. Furthermore, most speculations tend to focus on the on-screen relationship between "spectator" and the interface itself, rather than on intersubjective relations between participants within an environment that must eventually be left behind to return to embodied relations outside the matrix. Finally, as a consequence of this reductive and decontextualized understanding of identification, there have been continued calls for a studied detachment from visual pleasure, an unproductive tactic that serves only to cede the formidable affective power of visual identification to commercial interests. One might say that all three of the perspectives outlined above are fantasies of escape of one sort or another: The "iconoclastic" hopes to escape from the irrationality of the visual register into the relative "sanity" of the symbolic; the "happily narcissistic" hopes to escape the responsibilities of intersubjective relationships by relating primarily to its own inner baby; and the "cyborgian" perspective often seeks to escape from the confines of the fleshly, embodied world altogether. My fantasy, on the other hand, is that we use these emergent technologies to play with the *unheimlich*, this "ludic psychosis," not to escape our bodies "into the matrix" or to participate in narcissistic fantasies of the "I can contain all worlds" variety, and not to embrace a cyborg posthumanism, but rather to periodically estrange ourselves from our accustomed bodily parameters and the confines of the normative ego so that we may return "home" to our bodies and to our relations with other embodied creatures like ourselves with greater empathy and within a less tyrannical collective relation to the world of images in which we all must make our way.

Notes

1. "Work of the eyes is done, now/go and do heart's work/on all the images imprisoned within you; for you/overpowered them: but even now you don't know them."
2. As well as other characteristics, naturally. Here I am chiefly concerned with this phenomenon as it occurs in the visual register.
3. Examples here include a variety of input-output devices, including but not limited to electrode-embedded gloves (such as the DataGlove from VPL Research) or other garments that detect bodily motion and translate it into data, head-mounted displays, viewpoint-dependent imaging, and more exotic technologies currently under development such as retinal-scanning devices and "force-feedback" technologies that will translate visual data into tactile sensations.
4. This is a line of thinking that continues to draw explicitly on Bertoldt Brecht's notions regarding "epic theater." For a succinct overview of Brecht's major tenets, see Benjamin 1968.
5. The Lawnmower Man, you will be relieved to know, was redeemed by loving friendship in the end and returned to the real world. He was somehow physically crippled by his encounter with VR, but so much wiser in the end for abjuring its false promises of power.
6. The much-storied DataGlove, for example, which he developed at VPL and which was subsequently (and unsuccessfully) developed and distributed by Mattel as the PowerGlove in the first real attempt to market a VR accouterment to the general public. Currently, Lanier is working on VR technologies as head of the National Tele-Immersion Initiative.
7. I include here, with some reservations, Claudia Springer and N. Katherine Hayles.
8. In her more recent work, Hayles appears to have moderated her position, further emphasizing the grounding of being in embodiment. Her focus, however, is on an emergent form of "posthuman" embodiment that is distributed across a network of biological and technological nodes. See, for example, Hayles, 1999.
9. Not necessarily strongly determinative in the final instance, but determinative nonetheless, especially in terms of reception and use.
10. High-tech haruspication!
11. For a full definition of "cultural technologies" see Williams 1992.

12. There are accounts of this experience that aren't quite as literalized as Lacan's "mirror stage" that some might find more compelling. Regardless of whether there is a particular epiphanic moment in front of a silvered glass, however, the centrality of the self-image in subject formation remains the same.
13. Particularly in her *Computers as Theater* (1991), not surprisingly.
14. Unless, that is, the Extropians are right and we'll all be downloading our consciousness into the grid in forty years. Fortunately for us all, they are not.
15. "The Idoru" from HoriPro, Inc. Check her out on-line as featured in one of her many international fan pages!
16. Access is, as usual, an obvious but important caveat when the majority of the world's population still does not have a telephone.

Works Cited

Benjamin, Walter. 1968. "What Is Epic Theater?" In *Illuminations*, ed. Hannah Arendt, 147–154. New York: Schocken.

Boddy, William. 1994. "Archaeologies of the Electronic Vision and the Gendered Spectator." *Screen* 35, no. 2 (Summer):105–122.

Butler, Judith. 1993. *Bodies That Matter: On the Discursive Limits of "Sex."* London: Routledge.

Cubitt, Sean. 1996. "Supernatural Futures: Theses on Digital Aesthetics." In *Future Natural: Nature/Science/Culture*, ed. George Robertson et al., 237–255. London: Routledge.

Foster, Hal. 1996. "Obscene, Abject, Traumatic." *October* 78 (Fall):107–124.

Haraway, Donna. 1985. "A Manifesto for Cyborgs: Science, Technology, and Socialist Feminism in the 1980s." *Socialist Review* 80:65–108.

Hayles, N. Katherine. 1998. "Virtual Bodies and Flickering Signifiers." *October* 66 (Fall):69–91.

Hayles, N. Katherine. 1999. *How We Became Posthuman: Virtual Bodies in Cybernetics, Literature, and Informatics.* Chicago: University of Chicago Press.

Kipnis, Laura. 1993. *Ecstasy Unlimited: On Sex, Capital, Gender, and Aesthetics.* Minneapolis: University of Minnesota Press.

Lacan, Jacques. 1973. *The Four Fundamental Concepts of Psychoanalysis*, trans. Alan Sheridan. New York: Norton.

Lacan, Jacques. 1977. "The Mirror Stage as Formative of the Function of the I as Revealed in Psychoanalytic Experience." In *Écrits: A Selection*,. trans. and ed. Alan Sheridan, 1–7. London: Tavistock.

Lanier, Jaron. 1999. "Riding the Giant Worm to Saturn: Post-symbolic Communication in Virtual Reality." In *Ars Electronica: Facing the Future, A Survey of Two Decades*, ed. Timothy Druckrey with Ars Electronica, 242–243. Cambridge, MA: MIT Press.

Laurel, Brenda. 1991. *Computers as Theater.* Menlo Park, CA: Addison-Wesley.

Laurel, Brenda. 1992. "Brenda Laurel: Lizard Queen." *Mondo 2000* 7:83–89.

Marvin, Carolyn. 1988. *When Old Technologies Were New.* New York: Oxford University Press.

Mitchell, W. J. T. 1991. "Iconology and Ideology." In *Image and Ideology in Modern/PostModern Discourse*, ed. David B. Downing and Susan Bazargan, 321–330. Albany: State University of New York Press.

Mulvey, Laura. 1975. "Visual Pleasure and Narrative Cinema." *Screen* 16:7–18.

Pesce, Mark D. 1994. "Final Amputation: Pathogenic Ontology in Cyberspace." *SPEED: A Journal of Technology and Politics.* Available at: <http://www.hyperreal.org/~mpescu/Fa.html>.

Rheingold, Howard. 1991. *Virtual Reality.* New York: Simon and Schuster.

Rilke, Rainer Maria. 1980. *The Selected Poetry of Rainer Maria Rilke*, trans. Stephen Mitchell. New York: Vintage.

Silverman, Kaja. 1996. *The Threshold of the Visible World.* New York: Routledge.

Springer, Claudia. 1991. "The Pleasure of the Interface." *Screen* 32, no. 3:303–323.

Williams, Raymond. 1992. *Television, Technology and Cultural Form.* Hanover, NH: Wesleyan University Press.

| 12 |

ELOQUENT INTERFACES: HUMANITIES-BASED ANALYSIS IN THE AGE OF HYPERMEDIA

Ellen Strain and Gregory VanHoosier-Carey

At one time in the not so distant past, the intersection between computing and the humanities was inhabited almost exclusively by scholars in the areas of linguistic analysis, textual criticism, and archival work. These scholars reserved time on their campus mainframes to run programs that counted words or particular linguistic structures in a given body of texts and then used the resulting data to create concordances, compare versions of literary or historical texts, or reveal patterns spanning various works. The computer's role in this type of research was purely computational; the computer noted when and where particular words or patterns appeared and reported these as quantitative results—functions that could have been performed on any set of data. Although such instances constitute an application of computing to the humanities, they cannot be categorized as humanities-based computing. The utilization of computing capabilities, in such cases, was neither infused with nor based upon the precepts and goals that constitute the humanities.

For a given computing application to be considered humanities-based, the computer must provide a function integral to humanities work. This does not necessitate that the function be original, but it does require that the computer's functionality reflect, foster, or extend humanities thinking or learning. Based on this definition, true humanities-based computing could be said to have originated when writers and scholars first used the computer as an environment in which to explore humanities thinking as it relates to textuality, a moment otherwise known as the beginning of hypertext. By using the hyperlink's ability to facilitate multicursal navigation, humanities researchers and teachers could model effectively the postmodern theoretical concepts of textual contingency and instability (Landow 1992, 8–10; 128).[1] By providing this multicursality, the hyperlink made possible new writing environments such as *Intermedia* and *StorySpace* and, in turn, new types of texts that transformed these abstract concepts into processes that one could visualize and, more importantly, experience through the reading of said texts. Hypertextual novels such as Michael Joyce's *Afternoon* experientially foreground textual contingency by requiring the reader to make narrative choices; thereby, they virtually ensure that each reading of the novel is unique. Similarly someone who navigates a collaboratively written hypertextual web such as the *In Memoriam* project initiated by George Landow is practically confronted with the unstable, perspectival boundaries that separate text and context as well as author and reader.[2]

This chapter begins by revisiting this ur-moment of the hyperlink's emergence to better explain the current state of humanities-based computing. Over the last ten years, hypertext rapidly has given way to hypermedia. Linguistically, the switching out of "text" for the broader term "media" seems simple enough. Yet the de facto transition to hypermedia scholarship within the humanities is fraught with complexity and paradox. The adoption of hypermedia as a communication strategy suggests both a paradigmatic shift and a continuation of practices characteristic of the humanities. This shift, viewed in terms of its significance, involves an embrace of planes of discourse that, to date, have been underutilized within the dissemination of humanities scholarship. Rhetorical strategies take on an intricately faceted nature when hypermedia's visual, spatial, temporal, and interactive structures existing above and beyond textual argumentation are taken into account. Our particular perspective on these strategies—strategies we associate with the practice of design—has been honed by a series of focused efforts to translate humanities-based analysis into hypermedia forms. Although those efforts have taken place on various fronts, within this chapter we elucidate critical strategies of hypermedia argumentation and analysis with reference to *Griffith in Context: A Multimedia Exploration of* The Birth of a Nation *(GIC)*, an interactive application that we have designed over the past four years, most recently with the support of the National Endowment for the Humanities. *GIC* reveals the radically expanded contours of humanities-based rhetoric in the age of hypermedia as well as the conceptual and practical paradoxes associated with hypermediated scholarship. Although the shift in the position of humanities scholars from writers to designers inherent in the transition to hypermedia is admittedly quite radical, it also prompts a backward glance at long-standing humanities practices that reveals continuity as much as it suggests change.

Hypermedia as a Design Practice

To some degree, design entered the realm of humanities-based computing with the first visually marked hyperlink. The cues that advertise the hyperlink as a navigation opportunity and the functionality of that link, which transports one to some other mediated space, introduce another type of experience into the reading of text and of media. This experience is often conflated with the "reading" challenges with which the image confronts us. To interpret the revolutionary aspect of hypertext as the introduction of the iconic into text, however, is to miss the point. The visually marked hyperlink does more than provide another layer of signification through graphic means. Far more than acting as the author's yellow highlighter, it introduces a dimension of spatiality beyond that of the two-dimensional page. At the moment that the reader acts upon a link by clicking it with the mouse, the hyperlinked selection becomes more than a set of linguistic signs; it becomes a functional device acting as an interface rather than as image or text. The author assumes the position of engineer or architect as he or she designs navigability using the link as an entryway.

The resulting tension maps directly onto the term "hypertext" itself. The second half of the term, "text," clearly is rooted in the traditional humanities practices of textual analysis and written communication. The first half of the term, "hyper," provocatively suggests that the hypertextual form extends over and beyond these traditional textual practices. The hyperlink exceeds the boundaries of writing. As a design element, it cannot be accounted for in terms of textuality or language and demands to be situated within the context of design practice. Hyperlinking allows the written text associated with it to be more than a transcription of linguistic expression; it provides a means of constructing conceptual associations so as to create an environment for someone to explore interactively. Instead of relying almost exclusively on a writer's ability to denote his or her insights through language, the hypermedia author has a means of connotatively fostering and even demonstratively guiding such insight through the designed hypermedia space. Hypermedia's logic of purposefully constructed juxtaposition and association creates an architected meaning, an arena within which the demonstration of humanities methodologies can take on a dynamic form.

Unfortunately for the humanities, most of its scholars are not fully aware of the functional, design elements of hypermedia and the potential that they present to humanities scholarship. Instead, for them, hypermedia represents an experimental form of writing that challenges traditional notions of textual signification and roles of writer and reader—a prime example of Roland Barthes's concept of the "writerly" text whose broken surface reveals, among other things, the author's tangents and associational webs of thought. The reader becomes an active constructor of meaning, and the text becomes a set of navigational decisions that the reader must make (Barthes 1974, 4; Landow 1992, 5–6). Seen through this lens, hyperlinks represent breaks in the surface that reveal the indeterminate play of signifiers beneath; they are the tug on language's taut surface, the pull that initiates the fraying of meaning. With this celebratory positioning of hypertext as an unveiling of the ideological work of the signifier and of the authorial construction of rhetoric, hypertextual writing has been most often viewed as an absence or a lack of obfuscating gloss rather than the seed of another communicative mode. This emphasis on absence, on stripping away rather than constructing, has obscured the design elements of hypermedia as well as the conceptual and communicative avenues that design opens up for humanities-based analysis. Only by understanding hypermedia as a design practice can we fully realize the possibilities offered by humanities-based computing.

Hypermedia is especially suited for the humanities because of the strong affinity existing between it and humanities-oriented thinking and learning. In particular, it resonates with two characteristics of humanities expertise within a given domain: (1) a scholar's practical knowledge of the web of associations that link together a vast array of primary texts (written, visual, cinematic, material), secondary texts (historical studies, critical articles), and lived practices, and (2) a scholar's ability to discern patterns within and among these cultural artifacts to reveal such associations and articulate their significance within a set of culturally based narratives and practices. Hypermedia furnishes humanities scholars with new tools for modeling these associations and revealing their constructedness both for their own benefit and, more importantly, for the benefit of their students and the general public. A given hypermedia application provides those who interact with it an opportunity to explore a web of associations formulated by a scholar or set of scholars; moreover, it can serve as a means by which these users can construct their own such webs.

Interfaces and Design Communication

The resistance to greater usage of hypermedia within the humanities cannot solely or even chiefly be attributed to a long-standing ambivalence toward the multivalent power of the image. Although such ambivalence factors into this resistance, it obscures the role of a more crucial and more radical component of hypermedia, a design component that we have referred to as "interface" and vaguely associated with the concepts "spatiality" and "interactivity." Hypermedia interfaces are often largely visual, but they differ from images in that their role is primarily functional rather than representational; in other words, the visual interface is designed to be interactive, to produce, when manipulated, a particular change in the object or environment to which it belongs. Interface components convey their interactivity and invite manipulation by means of what James Gibson and others such as Donald Norman following his lead have termed an "affordance": a superficial quality of an object that conveys to a potential user by way of sensory information how one can interact with that object (Gibson 1977, 67–68; 1986, 127–128; Norman 1988, 9). For example, the finger holes of a bowling ball suggest by their size and relative orientation a particular way of handling the ball; similarly, the circular band on a set of standard headphones indicates by its relative position to the two ear-sized sound output devices that it should fit over one's head. Affordances, in short, convey the interactive significance of an object, the equivalent of meaning in words or images.

We use everyday consumer products here not as subjects for analysis but as examples of communication; by doing so, we underscore the radical paradigm shift that the demonstrative, interactive qualities of hypermedia entail and that humanities computing, to its detriment, has not yet undergone. Richard Buchanan (1989), in "Declaration by Design: Rhetoric, Argument, and Demonstration in Design Practice," attributes the difficulty of seeing the linkages between linear linguistic arguments and the arguments embodied and conveyed by consumer products to the mistaken belief "that technology is essentially a part of science following all the same necessities as nature and scientific reasoning" (93). The reality, however, is that product design, including the design of electronic applications, is more closely aligned with humanities work than with physical science. Buchanan asserts that product design, and technology more generally, are fundamentally concerned "with the contingencies of practical use and action" and thus participate in "the broader art of design, an art of thought and communication that can induce in others a wide range of beliefs about practical life" (93–94). In creating a consumer product, a designer delivers a demonstrative argument that initiates "an active engagement between designer and user" whenever that user "considers or uses [the] product as a means to some end" (95–96).

That applied arts are not only communicative but also argumentative is not some recent hypothesis; instead it is a notion rooted in Aristotelian ideas about art and rhetoric. As Buchanan (1995) explains in, "Rhetoric, Humanism, and Design," writing and applied forms of creative production were unified through Aristotle's notion of "forethought" as a master art that was concerned with discovery, invention, argument, and planning and was independent of specialization within a particular art. Reviving this notion of the common generative origins of applied art and writing in Aristotelian forethought, Buchanan argues for an understanding of the rhetorical features of an integrative "art of making" that embraces writing and design practices (31–32). Adopting Buchanan's stance, it is then possible to speak of design arguments that differ from written arguments only in their use of demonstration rather than direct assertion. That is, the designer conveys his or her argument through "a manipulation of the materials and processes of nature, not language," aiming to guide the user experientially toward a new attitude or mode of action (Buchanan 1989, 94).

Those particular elements involved in this interactive, experiential engagement comprise the product's interface, and they encourage desired applications of and attitudes about the product by means of their affordances. The mention of attitudes here is important because the product design's demonstrative argument is deemed successful to the degree that a product's affordances resonate with the user's perspective and thereby encourage the user to engage with the product. The visual aspect of these affordances is crucial, but it cannot be separated from function, ergonomics, tactility, and responsiveness; in turn, none of these qualities can be understood without taking into account the context in which the product is used and the assumptions of the user prior to this experience. The typical portable electric hair dryer, for example, functions as a tool to help the user efficiently dry and effectively style his or her hair. This function is made desirable by a cultural mindset and particular living situation in which time is scarce, personal control is valued, and stylish grooming is socially demanded. In this use context, a design that visually, spatially, and interactively conveys an understanding of and appreciation for these desires and expectations and thus promotes a set of appropriate attitudes while delivering the requisite function will successfully convey its argument. The successful dryer interface might include, among other things, an easily gripped handle that allows more freedom of movement, an air nozzle whose angle can be adjusted to facilitate styling, and power and temperature buttons that can be easily manipulated while the dryer is being held. More subtle features that, through responsiveness, suggest a general appreciation for the user might include buttons that visually announce their particular function and provide feedback through sound or mechanics of motion as they snap into place.

This discussion of product interfaces and affordances may seem initially quite distant from the pursuits of the humanities. For those in humanities computing, however, these design concepts are crucial, since they chart a future course that can extend humanities scholarship into the realm of application development while nonetheless remaining anchored to the core humanities practices of analysis, exegesis, and communication. The next generation of humanities computing applications will embrace design concepts and use them to create environments that will not only present humanities content but also, through careful attention to interfaces and the rhetorical power of their affordances, encourage the user to interact in ways that will promote particular analytical stances toward that content. Whether the design arguments of these applications can achieve the complexity and precision of their verbal counterparts is not really an issue, since there is no need for an either-or commitment when it comes to textual arguments versus design arguments. Textual arguments can and should be used when precise explanation is needed; this does not mean, however, that these humanities applications should be dominated by written arguments. We strongly believe that design arguments and the analytical stances that they can promote through nontextual means are crucial to the rhetorical success of hypermedia and future forms of humanities computing. Interactivity can be designed to produce rhetorical and pedagogical outcomes by modeling methodologies rather than merely asserting the end product of any particular methodology. With effective interface design, the user becomes a collaborator in the production of the argument in that he or she participates in the demonstrative act and utilizes the hypermedia environment as a phenomenarium in the active creation of knowledge.[3]

In the section that follows we discuss our own present humanities computing project *Griffith in Context* and use it as an opportunity to evince more concretely how interfaces and their affordances foster humanities-based analysis. Like the majority of humanities applications, *GIC* makes use of the rich interactivity of the hyperlink to facilitate its design argument; however, it also introduces specially designed interface objects that move beyond point-and-click navigation. Recognizing that the compelling interface of the early hypertext application *StorySpace* was due as much to its affordance of spatial mapping as to its facilitation of hypertextual connections, we have worked to extend hypermedia by attention to and exploitation of the space containing hyperlinks. More specifically, we have introduced a spatial coherence and sense of temporality that is crucial to fostering the close analysis and contextualized understanding of *The Birth of a Nation* that is our design objective. We hope that our design, in addition to being interactively compelling and pedagogically valuable, will prompt further experimentation with interfaces and affordances and thus new demonstrative arguments addressing humanities work.

Griffith in Context as Illustration of Humanities Design

To most eyes, *Griffith in Context* is a digital tool to aid in teaching. For us, it also serves as scholarship, both through its presentation of our scholarly insight into the film *The Birth of a Nation* and through its interface, which we custom-designed for the purposes of humanities-based analysis. Materially speaking, *GIC* is a dual CD-ROM computer application intended for use as part of a course or curriculum that is centered on cultural analysis from any of a number of disciplinary points of view: literature, film studies, American studies, history, and cultural studies. The application, intended for use by students outside of class, is dominated by a central module, which we will refer to as the main interface. This module features a hypermedia filmstrip with temporal annotations or links that provide entrance to scholarly voice-overs (SVOs) illustrated with archival material and contextualizing images.[4] Thirty-one scholars, each applying their areas of research expertise to the analysis of *Birth*, are featured on the CD. These SVOs, more than 100 of them in total, constitute the second module. The film-editing exercises, as the third and final module, round out the application as the most open-ended component in terms of the user's freedom to explore various alternatives.

We begin with our general teaching goals to show how we transformed into a designed artifact a set of pedagogical strategies intended to demonstratively convey an argument about the film's cultural significance. As cultural studies scholars, our primary interest in *The Birth of a Nation* was its ideological deployment of race, nation, class, gender, and regional identity, all confined within the benign guise of entertainment, and the cultural debates and discursive responses that the film mobilized. *Birth* is a perfect illustration of how cultural texts can be read as complex entanglements of ideological strands that must be teased apart to reveal the mechanisms of "cultural work." An important operative concept linked to this idea of ideological entanglement is Stuart Hall's notion of articulation, a process by which nonnecessary linkages of various conceptual and representational elements form historically situated discourses. Hall's use of the term "articulation" plays on the word's dual meaning: Cultural elements are articulated in the sense both that they are joined and that their cultural significance is expressed through particular cultural formations (Grossberg 1996, 141). From this perspective, *Birth* provides a set of textual moments in which certain cultural discourses became temporarily conjoined by way of visual representations, narrative constructions, and enacted cultural practices. These discursive embodiments resonated with the ideological tensions of the cultural and historical context in which the film was created and first viewed, thereby performing cultural work that had important ramifications for a burgeoning imperialist nation internally divided across regional and racial lines. Our goal for *Griffith in Context* has been to design an interface that fosters a way of seeing and interacting with the film that, to the degree possible, presents the film as articulated opportunities or "paths" that encourage the student users to explore the underlying discourses analytically. In other words, we have sought to make the experience of using the interface an integral part of our argument about the film. We attempt to achieve our pedagogical goals not through students' engaging with the content alone but instead through students' engaging with the content via the perspectives and user-initiated actions fostered by the interface. In this way, the content and the interface jointly and inseparably facilitate the application's pedagogical outcomes.

At this point, let us briefly return to Buchanan's claim about technological products discussed earlier. Buchanan (1989) argues that such products initiate a demonstrative argument about cultural concerns through the manner in which they address particular "contingencies of practical use and action" involving the user as he or she "considers or uses [the] product as a means to some end" (93–96). This assertion suggests that the design and implementation of an application like *GIC* clearly is a rhetorical problem. Our goal is to communicate certain ideas about Griffith's film to a specific audience in the most understandable and convincing manner possible. Doing so entails that we consider our audience's particular point of view and the assumptions and concerns that go with it and then focus our argument in a manner that addresses those assumptions and concerns. In terms of technology, our goal requires that we understand how our potential users typically engage with film and then devise a "tool" with technological affordances that address the contingencies these users most likely will face in analyzing *The Birth of a Nation*. We argue that most students approach film as entertainment. They expect a story to be visually conveyed to them and are neither aware of nor concerned about (1) the technical means by which film conveys meaning or (2) the film's meaning in terms of any historical or cultural context outside of their own personal experience. This is not to suggest that their spectatorial skills are somehow deficient; on the contrary, a director such as Griffith fully expected such a stance and actually depended on it to create the naturalized sense of reality—the sense of history come alive—for which he aimed. Yet this spectatorial stance makes teaching Griffith's film quite challenging, since the technical means and the early-twentieth-century historical and cultural contexts are primarily what most professors teaching the film want their students to focus on and analytically probe. Thus, any tool designed to foster an analytical stance toward such a film must encourage a defamiliarized viewing that underscores filmic technique and contextualized meaning.

GIC attempts to do this not by jettisoning the spectatorial viewpoint, but instead by providing an interactive, analytically geared apparatus around it. In other words, *GIC* allows the student to view clips from the film in a traditional manner but within an interface that encourages the student to approach the film from other perspectives that highlight its technical and cultural constructedness. The main interface's navigable filmstrip first confronts the student with a defamiliarized perspective on the film (figure 12.1). The navigable filmstrip presents the film as a spatially ordered series of still images similar to the celluloid strip that passes through a projector. In so doing, the spatialized "filmstrip-view mode" breaks the naturalizing effect of persistence of vision, which encourages the eye to see the still film frames as a continuous and moving image, thus revealing motion to be the film's central illusion. The spatialization is

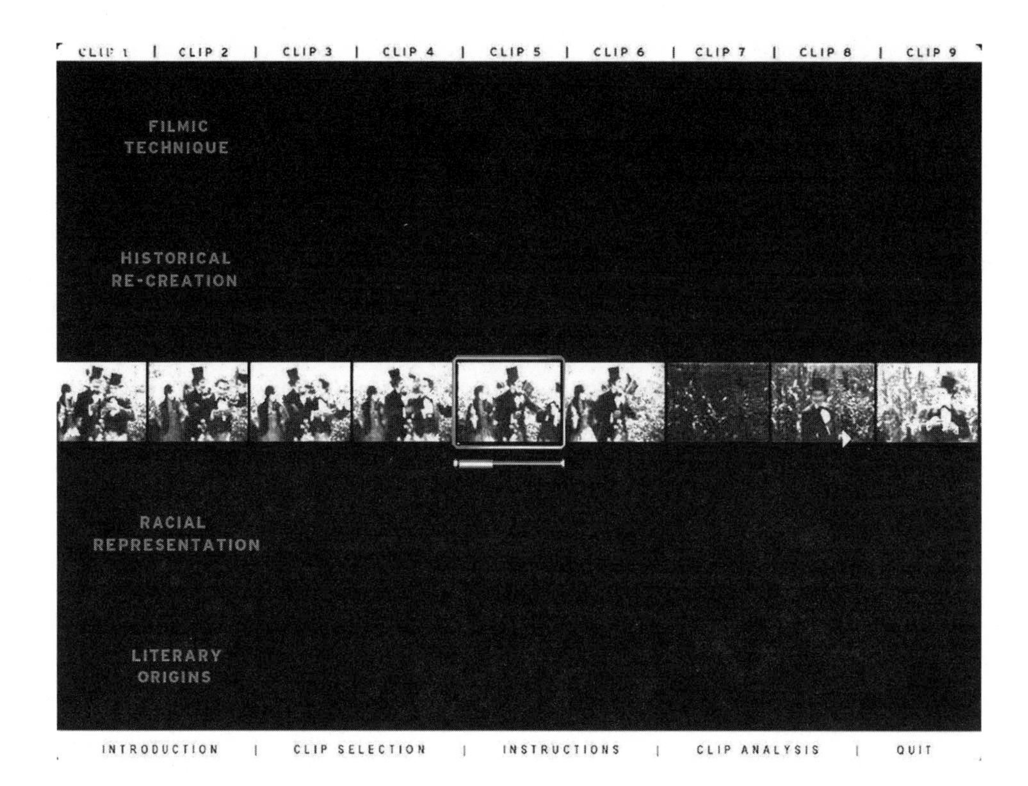

| Figure 12.1 |

GIC's filmstrip-view mode offers a spatialized navigable filmstrip that foregrounds the film's constructedness.

underscored by the navigable filmstrip's affordances and functionality. The extension of the filmstrip images off-screen prompts the user to "pan" to the right to explore the film's continuation in the same way that an activation of off-screen space often motivates a horizontal camera pan in film. The filmstrip's advancement in response to this mouse-controlled pan confirms the film's status as a spatial rather than temporal object and thus furthers the revelation of the film's defamiliarization and the constructedness it suggests.[5]

The ability to manipulate the navigable filmstrip also is significant because it simulates the point of view of a film editor handling the celluloid strip. The main interface's condensed view of the film, representing sixteen actual film frames with a single frame on the annotated filmstrip, contributes to this editorial perspective by emphasizing cuts and disjuncture rather than continuity and similarity. Working from a knowledge of Griffith's average shot length, we recognized that a filmstrip view representing sixteen still images per second would result in a single shot spanning multiple computer screen widths. The student would be drawn to the small differences between frames, differences that, when stitched together with a sweep of the eyes across the screen, would focus attention on action within the frame. The condensed filmstrip with one frame representing each second of the film, on the other hand, draws the eyes to the cuts, that is, the seams that remind us of the film's constructedness.

Our design of the navigable filmstrip thus builds upon the constructivist model of education and on a particular vein of political analysis within film theory. The learner is encouraged to play an active role by the very fact that the filmstrip remains static without the student's participation. The speed and direction of filmstrip movement are both controlled by the student's mouse position. In addition to inviting participation, the filmstrip starts to reverse the obfuscation of cinema's ideological work, which begins with cinema's reliance on the persistence of vision to hide film's nature as a series of static images. This illusion of motion is only the most recessed of multiple levels of obfuscation in operation within film. At another level, continuity editing—a core component of Hollywood cinema's tendency toward invisible storytelling—obscures the essential disruption of the cut by providing a strong through-line from one shot to the next, rendering the edit nearly invisible to the engaged spectator.

GIC's navigable filmstrip brings both of these illusions back into view while connecting them to the naturalizing function of narrative. Attention is refocused from the mechanics of film's composition to the narrative itself by way of a toggle function that allows the student to move back and forth between filmstrip-view mode and traditional viewing of the clip (clip-view mode) starting from the central film frame within filmstrip-view mode (figure 12.2). Understanding *Birth*'s cultural work requires both this unveiling of the film's labor at the level of individual shots stitched together to create meaning and a reminder of the viewing experience (i.e., the power of the film when experienced as a kinetic narrative brought to life on screen). The application's facilitation of the student's swift movement between filmstrip-view mode and clip-view mode facilitates this kind of mobility in thought, an ability to connect issues of the film's construction to the power of the narrative. *Birth*'s ideological articulations become visible through *GIC*'s own articulations: its ability to join issues of form, such as the juxtaposition of shots, to issues of content, such as race, and to express or articulate through design how the film performs cultural work.

This ability to toggle between a viewing mode that supports narrative immersion reminiscent of the spectatorial experience and an analytical mode that foregrounds the film's construction is a "bifocal" approach that is echoed in the use of scholarly

———

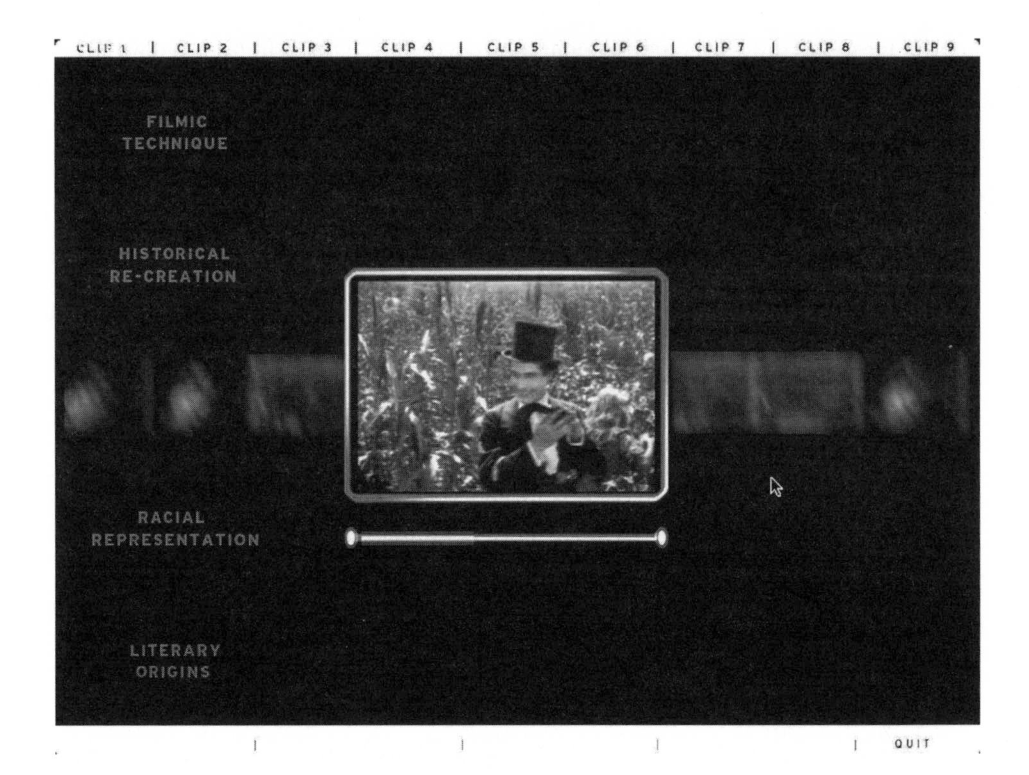

| Figure 12.2 |

The user can toggle into clip-view mode and begin playing a clip by clicking on the central film frame in the navigable filmstrip.

voice-over (SVO) links. The filmstrip-view mode moves the student outside the film's diegesis. Similarly, the SVO links appearing above or below the band of the screen utilized by filmstrip-view or clip-view mode draw the eye momentarily outside the film (figure 12.3). Whether a clip is being viewed through traditional playback or analyzed via the filmstrip, links to SVOs appear in the margins timed to the appearance of certain images within the clip. These temporally juxtaposed links highlight moments of ideological articulation or pathways into discursive strands associated with the clip. For instance, the exchange of the cotton blossom for a photograph of Elsie, a montage sequence from the film featured in *GIC*'s first clip, might be accompanied by a clustering of SVO links that pop up at this point in the application. The clustering of these links marks the sequence as a moment of articulation that prompts comment by a number of scholars. The nature of articulation and the moment's significance are then

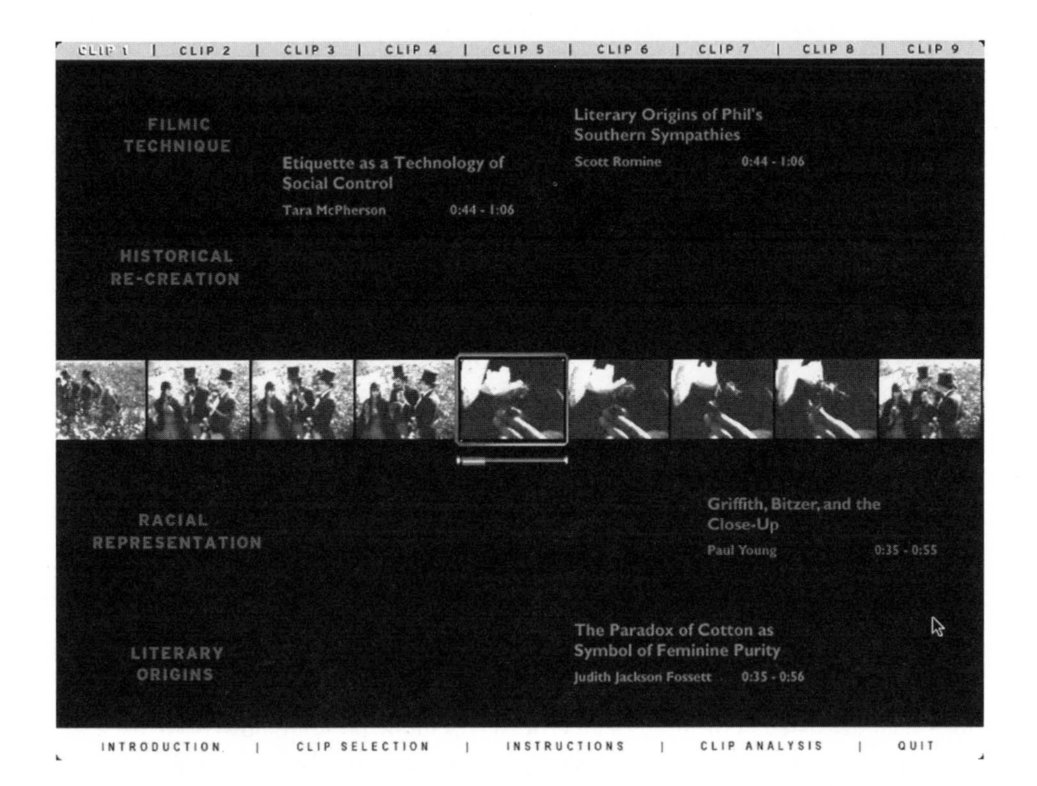

| Figure 12.3 |

Links to SVOs are temporally juxtaposed with particular frames in a film clip. This juxtaposition highlights moments of ideological articulation or pathways into discursive strands associated with the clip.

further visually revealed through a color-coding schema. Each SVO is categorized in terms of four indices of analysis (Filmic Technique, Historical Re-creation, Racial Representation, and Literary Origins). Thus, when a cluster of links appears, their colors visually suggest something about the cultural significance of representative elements. A cluster of green links appearing would suggest that the textual moment featured at that point in the clip "leads into" discursive issues most prominently related to historical concerns; a cluster of different-colored links, in contrast, would suggest a textual moment in which issues of historical depiction, literary tropes, cinematic techniques, and constructions of races intersect. The titles of the individual links provide still further information about the moment's significance, information that then is more fully elaborated on in the SVO discussions to which they lead.

Clicking on an SVO link brings the student user to a new screen, one that no longer features the filmstrip at its center. The removal of the filmstrip at this point represents a temporary reorientation for the user. The clip-view and filmstrip-view modes of the main interface and the "perspective toggling" that the existence of the two modes facilitates encourages an exploration of the cinematic constructedness of the clips. The SVOs, on the other hand, promote an analytical perspective focused on contextualized meaning. This perspective builds on the representational and ideological cues conveyed by the clustering and color coding of the SVO links. Having explored a clip and become interested in a moment of articulation, the user enters a space in which he or she has the opportunity to learn about that moment in more detail. This learning opportunity is divided into two phases. The first phase involves a structured presentation in which the user listens to a scholar analyze a given theme or issue arising from a particular aspect of a clip. These audio analyses are coordinated with a visual presentation of film stills as well as graphic artifacts and primary documents directly and indirectly associated with the analyses (figure 12.4). Once the presentation is completed, the user is free to explore in a less structured manner. The presentation's various documents become available for the user to move around the screen and examine more closely. The user might choose, for example, to read a review of the film or a court document discussed in the presentation or to reexamine photographs.

The very ability to "handle" primary documents is suggestive of the nature of scholarly work, thereby providing a glimpse into humanities research as conducted by the experts featured in *GIC*'s SVOs. Additionally, the movement of documents made possible by the interface provides opportunities for the user to create new juxtapositions, fresh readings, and new discoveries by comparing film frames, for instance, to early-twentieth-century advertisements or scientific illustrations. This revelation of the nature of scholarly work provides a parallel to the interface's revelation of the cultural work of the film itself. Similarly, the potential to generate new scholarly insight by enabling access to the archival materials provides a parallel to the opportunity to create new filmic sequences facilitated by the editing exercises, as discussed below. It is also significant to note that the SVOs are not overly polished narrations, but rather scholars' talking through their thoughts in a dynamic style that captures the excitement of fashioning new conclusions as they use their established research to view *Birth* in innovative ways. Equally important, the SVOs reverse the traditional tactic of beginning an argument with a thesis and then presenting only those details that buttress that thesis. The SVOs convey the idea that arguments begin as details of a text that evolve into compelling evidence in support of an argumentative claim rather than foster the illusion that argu-

| **Figure 12.4** |

Each SVO features an audio analysis coordinated with a visual presentation of film stills, graphic artifacts, explanatory notes, and primary documents. Once the analysis is completed, the user can explore the visual artifacts independently.

ments emerge fully formed from a scholar's head. The end result is an introduction to humanities argumentation that also provides archival materials; the structured presentation in combination with document access provides contextual knowledge and models analytical assumptions and methods in ways that encourage the user to utilize these analytical models further. Students are also encouraged to explore themes introduced in individual SVOs by way of thematic links or "theme cards" that tie together SVOs across the application (figure 12.5). Thus, although the SVOs are grounded in and linked to particular filmic moments, this webbed structure of thematic links facilitates the tracing out of themes as they resonate across the film.

Taken collectively, the SVOs can be seen as a multifaceted viewing apparatus, much like a gem whose refracting surfaces bring new perspectives on the film. Distinct

| **Figure 12.5** |

Thematic links or "theme cards" tie together SVOs from across the application, encouraging *GIC* users to explore themes introduced in individual SVO.

facets are provided by our scholars, who are identified by name, voice, photograph, and institution within each SVO and who are grouped by interest through the color coding of the link to their SVO. The array of SVOs foregrounds diverse ways to examine the film that demonstrate scholars' distinct intellectual approaches applied to the same object of analysis. The relationship between these approaches is depicted in the interface as nonhierarchical and multicursal with rich areas of intersection, complementary discourse, and even productive dissension indicative of debates within and across disciplines. They encourage an interdisciplinary perspective that nonetheless acknowledges disciplinary differences and thus can serve the needs of instructors and students from a variety of academic fields, including film studies, American studies, African American studies, and cultural studies.

A small number of links in the filmstrip-view mode of the main interface lead not to SVOs but instead to editing exercises that encourage the student to reedit a particular scene from *The Birth of a Nation* in ways designed to alter the scene's meaning and ideological resonance. The editing exercises facilitate complete manipulation of a given clip by allowing the user to cut and rearrange one-second film fragments in addition to altering sound track, tints, and masks. Each such exercise focuses on a particular technique featured in the clip where the link appears and discussed in the filmic-technique SVOs. For example, the editing exercise for the third clip in *GIC* focuses on parallel editing, a concept discussed in the SVOs and exhibited in the clip itself. Having been exposed to the concept, the user can then use the editing exercise as a place to apply the technique and test the claims made in these SVOs. Because each exercise uses the same basic interface with the same set of affordances, the user is not restricted to experimenting solely with the technique featured in any particular exercise. After having viewed other clips and associated SVOs, he or she might experiment with multiple techniques: those featured in the current clip as well as previously encountered clips. For example, in the parallel-editing exercise, the user might also try applying different-sized masks to sets of frames to focus attention on one part of the shot; the user similarly might alter the meaning of the clip by using a different soundtrack or applying a color tint to the footage. In other words, the student is given access to some of the same tools that Griffith himself used to suggest contrast or parallels between characters or actions, characters or actions that stand in for larger discourses within the film.

The design of the editing engine interface plays a crucial role in making this experimentation pedagogically valuable. The employment of particular editing techniques is not the critical component of the editing exercises. The pedagogical value stems instead from the editing exercises' encouragement of specific thinking about how cinematic narrative conveys meaning. The ability to reedit sequences makes tangible what many of the scholars in the SVOs argue, that is, the interconnections of filmic technique and ideological articulations. Users, in fact, can reweave ideological articulations by creating new filmic sequences from Griffith's footage. The sophisticated functionality of the editing exercises is made familiar through the use of a modified version of the filmstrip featured in the main interface. The user navigates through the strip to find the beginning and ending frames of the first cut; at those specific frames he or she inserts an in point and out point, respectively (figure 12.6). Once these points are selected, a copy of the initial frame appears in the first of eight "bins" arranged in a row from left to right just below the filmstrip; the next bin then becomes

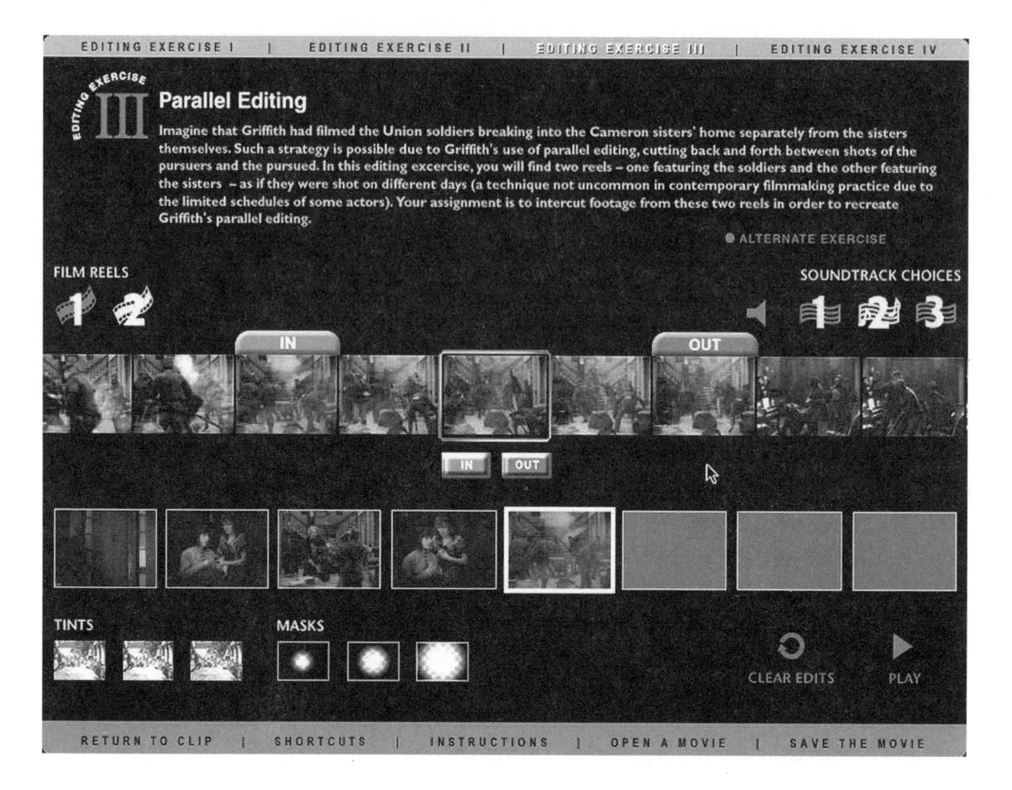

| Figure 12.6 |

Editing exercises allow users to reedit portions of the film clips as well as add tints, masks, and various soundtracks. These exercises foreground the role that editing and other means of technical manipulation play in constructing meaning.

automatically selected, enabling the user to make another cut by inserting in and out points. A series of cuts thus is represented in the editing interface as a progression of initial frames, each marking a shot; this representation highlights transitions between shots, implicitly suggesting that such transitions and their order are crucial to the narrative. The progression of images across the bins reinforces a logic of juxtaposition as students create their own filmstrips. In the process, the original filmstrip and its meanings no longer appear inevitable; instead, filmic meaning is viewed as highly plastic.

The interface also supports learning by encouraging revision and further experimentation. Cuts may be moved within the edited sequence; for example, one can move the second cut to the fourth position by dragging the contents of the second bin into the fourth bin. The user also can make changes to a given cut by clicking on the corre-

| Figure 12.7 |

After editing a sequence, the user can view the finished product; the user then has the opportunity to make additional changes to the sequence.

sponding bin. This action brings up that cut's in and out points on the filmstrip, allowing the student to alter these points or to use the mask and tint features to change the look of the new shot. When the user is finished editing a sequence, he or she clicks on the "premiere movie" button, which takes the user to the "premiere" screen, a theater environment in which the newly edited sequence is played (figure 12.7). After the sequence finishes, the user is returned to the editing interface, where he or she can make further changes. This separation of editing and viewing into separate spaces with very different affordances is meant to convey to the user that successful editing requires alternating between the point of view of editor and that of the audience. Our goal is to underscore this necessary shift in perspectives through a corresponding spatial shift.

Evaluation of these reedited sequences using the narrative and formal principles introduced by *GIC* can take place iteratively and collectively. A student may save the edited sequence to his or her own computer, return to his or her prior work after the passage of time, and then reconstruct the sequence using knowledge accumulated since this prior use of the particular editing exercise. The ability to try another approach and to create multiple edited sequences from the same material points out the malleability of the film form. Additionally, the portability of the saved sequences means that edited sequences can be exchanged between users or between teachers and students who have copies of *GIC*. Through such exchanges, one can glimpse each student's interests, agendas, and constructions of knowledge in connection with the cultural issues raised by the film; the sharing of such perspectives results in a widened knowledge base for all participants. Lastly, the sharpening of visual communication skills through strategic manipulation of computer-based images helps undergraduates make the transition to digital- and screen-based rhetoric, with which so many text-based scholars are currently grappling. In short, we use design to train students to understand the image and to be designers themselves in a world in which communication encompasses far more than language alone.

Conclusion: Design Practice and Its Place in Humanities Work

The majority of scholars and instructors who have viewed *Griffith in Context* have expressed enthusiasm for it, enthusiasm not only about having an electronic application that can assist them in teaching Griffith's problematic yet important film but also about the means by which *GIC* accomplishes this. In other words, these scholars and instructors have perceived a "fit" between the design of *GIC* and its content and pedagogical goals. This reaction, along with positive results from the initial classroom testing of *GIC*, suggests that the application can be an effective humanities learning tool; it also validates our conviction that the future success of humanities-based computing depends upon our willingness and ability to bring humanities methods and concepts to bear on the interactivity of humanities-based applications. We argue that humanities-based computing should be a widening and deepening strain within the academy and that the most promising approach is for humanities scholars to become intimately involved in the design and implementation of humanities-based applications. To do this, we need to acknowledge the continuity between design efforts and the long-standing traditions of humanities work and to understand the practical initiatives of humanities-based computing as legitimate scholarly work.

Humanities scholars' present lack of awareness concerning design is evident in the limited integration of hypermedia into current scholarly work. Rather than rethinking one's communicative goals in terms of design's multiple planes of communication and then crafting an experiential object for a particular use context, the majority of scholars have instead "disciplined" hypermedia by requiring it to adapt to traditional modes of scholarly argument. The strategies of intellectual taming applied to hypermedia's multicursality are in many ways parallel to those applied to the image, another threateningly connotative media form. Graphic communication has been shunned in favor of the scholarly translation of image-based culture into familiarly textual forms; similarly, hypermedia's multicursality and multimedia forms have been a subject of linear textual analysis without significantly influencing the communicative modes of humanities scholarship.[6]

The irony of this stance is striking: Humanities scholars embrace hypermedia because it encourages reader participation in the text's construction of meaning and models the associative connections within and between texts obscured by print technologies yet fail to take advantage of these qualities by implementing this mode within their scholarship. In other words, that very aspect of hypermedia which new media scholars have found so intriguing—its participatory yielding of control to the reader—appears to run counter to the traditions of humanities scholarship. In contradistinction to the assumed collaborative creation of meaning within hypermedia, the publishing tradition within the humanities foregrounds the scholar's voice in a solo performance of intellectual authority. The ceding of even a small part of this authority to another, as in the case of hypermedia navigation, seems to raise doubts about the scholar's virtuosity in regard to his or her subject. Such a perspective would be logical within a profession whose activities are exclusively solitary in nature; however, such is not the case with humanities scholars. In fact, the communicative task that shares near equal billing with scholarly publication within most academics' careers—teaching—is one that is quite interactive. Essentially, humanities teaching is a design task in which the instructor creates a set of experiential learning opportunities associated with humanities work and then guides student participation in them.

One clearly can see the design strategies involved in teaching by examining the strategies and decisions involved in creating a course syllabus. Ostensibly, the purpose of a syllabus is to communicate a schedule of readings and assignments to students. More importantly, however, the syllabus document maps out a particular path or set of paths through a body of knowledge and demarcates the experiential learning opportunities in the form of class lectures or discussions on particular readings or topics. One of the first decisions a teacher of a course must make is the conceptual schema for the course. Will an American literature course, for example, be organized thematically or historically? This decision will help determine the overall shape or orientation that the subject matter will take as the student moves through the readings; it also will encourage the student to make particular connections with his or her experience and play down other connections. The choice and order of readings is another crucial design decision. The selection of one particular text instead of another for the initial course reading may set an entirely different tone for the course or put in play a particular set of concerns that, in turn, will set the stage for later readings. The scheduling of projects or exams is similarly crucial, since what has been covered in class will determine what can be assessed in the first exam or project and thus what must be assessed in subsequent assignments.

These are only a few of the design challenges associated with teaching, but they are enough to demonstrate an important point: the design task of humanities teaching may involve traditional unicursal argumentation, but it is not reducible to it. Although writing lectures or discussion notes is an important part of teaching, we must not forget that it is the deployment of the lecture along with demonstration materials, questions to the students, opportunities for students to ask questions, and other activities—all at a planned point in the course—that fosters learning. In other words, humanities teaching is the crafting of learning experiences, the construction of a scaffolding and a set of activities that model humanities methods and assist students in acquiring them through guided practice. The activity of crafting becomes even clearer when one takes into account more contemporary learning paradigms such as those advocated by constructivist educational theories. These theories place greater emphasis on student participation than on the dissemination of empirical knowledge and thus advocate a learning environment within which the student pursues and tests his or her own hypotheses, integrating conclusions into his or her outlook on the world. This school of teaching exchanges the lecture for active learning tasks, but it does not eliminate analysis and argumentation; instead what occurs is that the teacher more completely integrates these essential humanities processes into the learning experiences. The analytical argument is no longer explicitly stated unicursal writing but instead is a set of implicitly conveyed interactive experiences designed and guided by the instructor in a manner that fosters a set of pedagogical goals.

We argue that the distinctions between the two humanities communication situations, teaching and scholarship, rest not in essential differences between teacher/student and author/reader relations but in the traditional mode of communication for each of these activities, paper-based textual forms versus the interactive environment of the classroom. Examples of humanities computing such as hypermedia encourage the collapse of such distinctions and allow design strategies analogous to those of teaching to be employed within humanities scholarship. Viewed in this light, what commonly has been seen as the yielding of authorial control within hypermedia forms can be understood as the insertion of greater effectiveness through a planned experience that encourages greater participation. The experience of the reader or user can be shaped by the hypermedia creator in the same way that students' learning experiences are sculpted by effective teachers. Common fears that the participatory experiences allowed for by hypermedia forms will not involve the same level of control over the communicative event stem not from the deficiencies of interactive forms but from a lack of understanding of the multiple planes for argumentation within hypermedia. In short, humanities scholars locked within a self-image that defines them as writers have failed to transition to an understanding of their role as designers, a role that would bring greater effectiveness to their scholarship and an easing of the artificial divide that separates research and teaching within the humanities.

Design practice, both the iterative process of developing a design and the direct engagement with the technology necessary to implement that design, provide opportunities to develop and apply humanities concepts in much the same way that scholarly writing does; the major difference is that the mode in which one articulates these concepts is a practical, demonstrative one. This claim calls into question the idea that scholars who practice humanities-based computing are a "hybrid breed" of scholars or not scholars at all because they also devote time and energy to learning technology and staying abreast of its changes. From the perspective of design practice, a humanities scholar's proficiency in understanding and implementing technology is analogous to his or her proficiency in analyzing and producing effective written text; in both cases, such proficiency involves engaging with prior arguments about a given set of concepts and issues and producing new arguments that help to clarify those concepts and issues further. As we have shown above using the insights of Richard Buchanan, technological products are rhetorical devices that present demonstrative arguments. They articulate, in both senses of that term, practical meaning that shapes attitudes and encourages future action through interactive engagement with users. Addressing cultural articulations through the demonstrative mode of design practice provides new perspectives on the ideological effect of these articulations, perspectives that are conveyed experientially rather than textually. This mode is one that we cannot afford to dismiss in an age in which information and insight increasingly are conveyed in nontextual form. Doing so would mean isolating the humanities from contemporary concerns at a time when insight from the humanities is needed most.

Notes

1. Throughout his discussion in *Hypertext*, Landow (1992) refers to the hyperlink as a means of multisequential navigation; here and throughout this chapter we have replaced the term "multisequential" with "multicursal." In doing so, we follow the lead of Espen Aarseth (1997), who has persuasively demonstrated the advantages of using the terms "multicursal" and "unicursal" to distinguish between textual environments that have a network of possible pathways and those with only one possible path (43–44).
2. For a good discussion of Joyce's literary application of hypertextual features in *Afternoon, a story*, see Bolter 1991(123–126). Landow (1992) discusses the *In Memoriam* project (141–145).
3. D. N. Perkins (1992) defines a phenomenarium as "an area for the specific purpose of presenting phenomena and making them accessible to scrutiny and manipulation" (47). Hypermedia applications such as *GIC* qualify as phenomenaria to the extent that their logic of juxtaposition and contextualization allows users to explore the complex dynamics of intertextuality.
4. Both the term "temporal link" and the idea for using time-based navigation in our application come from Sawhney, Balcom, and Smith's work on hypervideo (Sawhney, Balcom, and Smith 1996, 3–4; 1997, 32–34). For a working demonstration of hypervideo, see the HyperCafe Web site at <http://www.lcc.gatech.edu/gallery/hypercafe/>.
5. The interactivity that we use in the filmstrip-view mode is quite similar to that used in Apple's *QuicktimeVR* application; the ends, however, are quite different. *QuickTimeVR* attempts to orient one to a space by allowing one to see 360 degrees, as if one were there. *GIC*'s filmstrip-view mode, in contrast, aims to defamiliarize rather than simulate the actual experience of the film viewer; it attempts to spatialize an experience that is not normally understood as spatial. More information as well as links to examples of *QuickTimeVR* is available on Apple's *QuickTimeVR* Web site at <http://www.apple.com/quicktime/qtvr/>.
6. The few examples of professionally successful scholarship in hypermedia form fall neatly into two camps, neither of which uses hypermedia's design affordances to conduct humanities-based analysis. The first camp consists of articles in online journals or hypermedia ancillary materials that use hyperlinks for footnotes, glossary entries, and transitions to subsequent sections; these works do not treat the hypermedia environment as a communicative mode that offers unique means of demonstrating points and shaping attitudes. The second camp, which includes most hypertextual fiction, incorporates the spatial network allowed for by the hyperlink but focuses on textual and multimedia experimentation rather than on promoting humanities-based analysis and argumentation.

Works Cited

Aarseth, Espen. 1997. *Cybertext: Perspectives on Ergodic Literature.* Baltimore: Johns Hopkins University Press.

Barthes, Roland. 1974. *S/Z*, trans. Richard Miller. New York: Noonday.

Bolter, J. David. 1991. *Writing Space: The Computer, Hypertext, and the History of Writing.* Hillsdale, NJ: Lawrence Erlbaum.

Buchanan, Richard. 1989. "Declaration by Design: Rhetoric, Argument, and Demonstration in Design Practice." In *Design Discourse: History, Theory, Criticism*, ed. Victor Margolin, 91–109. Chicago: University of Chicago Press.

Buchanan, Richard. 1995. "Rhetoric, Humanism, and Design." In *Discovering Design: Explorations in Design Studies*, ed. Richard Buchanan and Victor Margolin, 23–66. Chicago: University of Chicago Press.

Gibson, James J. 1977. "The Theory of Affordances." In *Perceiving, Acting, and Knowing: Toward an Ecological Psychology*, ed. Robert Shaw and John Bransford, 67–82. Hillsdale, NJ: Lawrence Erlbaum.

Gibson, James J. 1986. *The Ecological Approach to Visual Perception.* Hillsdale, NJ: Lawrence Erlbaum.

Grossberg, Lawrence. 1996. "On Postmodernism and Articulation: An Interview with Stuart Hall," In *Stuart Hall: Critical Dialogues in Cultural Studies*, ed. David Morely and Kuan-Hsing Chen, 131–150. London: Routledge.

Landow, George P. 1992. *Hypertext: The Convergence of Contemporary Critical Theory and Technology.* Baltimore: Johns Hopkins University Press.

Norman, Donald A. 1988. *The Design of Everyday Things.* New York: Doubleday.

Perkins, D. N. 1992. "Technology Meets Constructivism: Do They Make a Marriage?" In *Constructivism and the Technology of Instruction: A Conversation*, ed. Thomas M. Duffy and David H. Jonassen, 45–56. Hillsdale, NJ: Lawrence Erlbaum.

Sawhney, Nitin, David Balcom, and Ian Smith. 1996. "HyperCafe: Narrative and Aesthetic Properties of Hypervideo." In *Hypertext '96, Washington, D.C., March 16–20, 1996: Seventh ACM Conference on Hypertext: Proceedings*, 1–10. New York: Association for Computing Machinery.

Sawhney, Nitin, David Balcom, and Ian Smith. 1997. "Authoring and Navigating Video in Space and Time." *IEEE Multimedia* 4, no. 4:30–39.

| 13 |
WRITING A STORY IN VIRTUAL REALITY

Josephine Anstey

Introduction

On the outskirts of new media, virtual reality (VR) is still lurking. Announcements of its demise (following the deluge of criticism and disappointment that followed the hype that starred *you* in a customized action/adventure/romantic movie in the Holodeck in your family room) were exaggerated. What remains are the challenge, the potential, and the limitations of making meaning using technology that can plunge the reader-user into a three-dimensional audiovisual world that responds in real time to her interventions.

This chapter is primarily an account of my experiences writing *The Thing Growing*, an interactive fiction for immersive VR. Using the work of Brenda Laurel and Janet Murray, Michael Mateas (2001) defines one "niche" in the range of possibilities for interactive narrative in which the user assumes the role of a first-person character and the Aristotelian dramatic strategies of enactment, intensification, and unity of action are used (51). I consider *The Thing Growing* to fit into this interactive drama "niche."

The chapter's first section describes the VR system I use for my work and the *The Thing Growing* application. The second section discusses the break with and extension of traditional writing methods that writing for VR entails. It is composed of three main areas: the place of visual design in the authoring process; how writing computer code is and is not like writing prose; and issues of user control in creating interactive fiction in VR.

The CAVE

The Thing Growing was originally built for a CAVE VR system (Cruz-Neira et al. 1992, 65–72). The CAVE (figure 13.1) is a room-sized, projection-based, virtual reality

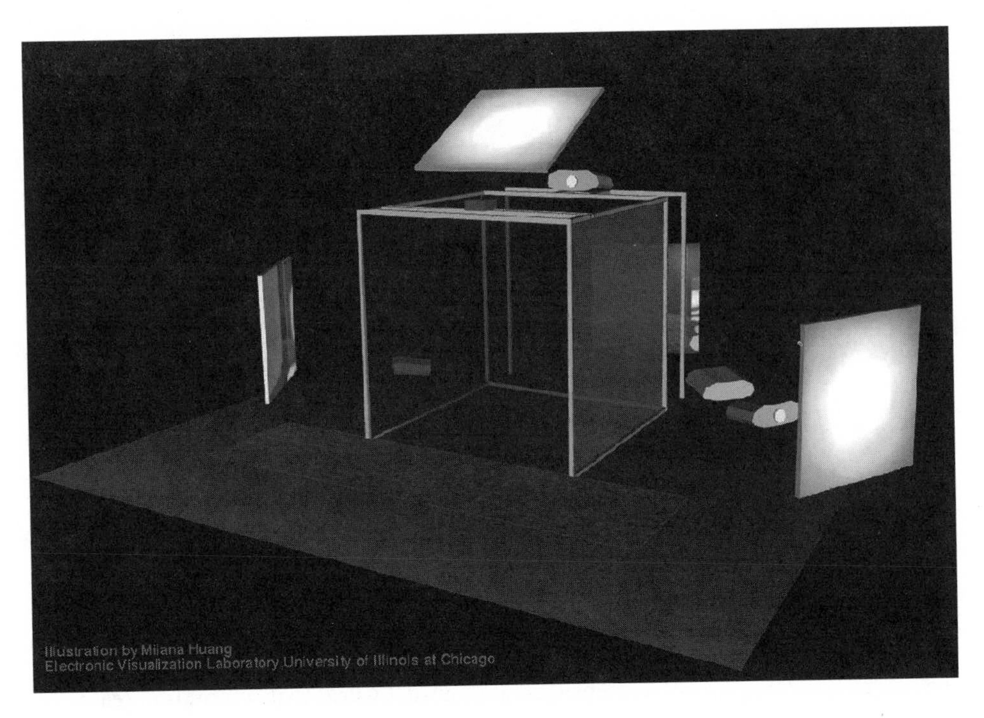

| **Figure 13.1** |

The CAVE, by Milana Huang. Courtesy Electronic Visualization Laboratory, University of Illinois at Chicago.

theater developed at the Electronic Visualization Laboratory (EVL) at the University of Illinois at Chicago in 1992. Projection-based VR is exactly the same as head-mounted VR, except the user doesn't have to carry the display equipment in the helmet on her head. The same mathematics and depth cues are given to immerse the user in a three-dimensional virtual environment. To see the virtual environment in stereoscopic 3D, the user wears active stereo glasses. A very high speed graphics computer displays an image for the right eye and the left eye sequentially. The glasses are synchronized with this process and alternately block the left and right eye. The user's brain puts the two images together and creates a stereo image. The CAVE system includes a wand equipped with a joystick for navigation and three buttons for interaction. The user has two distinct ways of moving in the space. She can move her physical body, walking about within the confines of the CAVE's walls, crouching down to look under 3-D objects that occupy the space. Or she can use the joystick to navigate anywhere in the virtual environment; in this case it is as if she were driving the CAVE through the virtual environment. In both cases the computer draws the graphics from her current point of view. Tracking sensors are attached to the user's body to feed information about her position and orientation to the computer. Typically in the CAVE the user's head and one hand are tracked. For *The Thing Growing* project I added a tracker for the other hand. In projected VR systems like the CAVE, the user sees her own body in the 3-D environment and automatically uses her own size to judge the scale of objects in that environment.

The Thing Growing

The goal of *The Thing Growing* is to make the user the protagonist of an interactive story. My original purpose was to describe a claustrophobic but emotionally hard-to-escape relationship. The original medium was a short story; readers would identify with the protagonist's feelings as she experienced the relationship. When I moved the story to VR, I planned to create the same tensions and emotions directly in the user, who is lured into a relationship with the Thing (figure 13.2), a computer-controlled

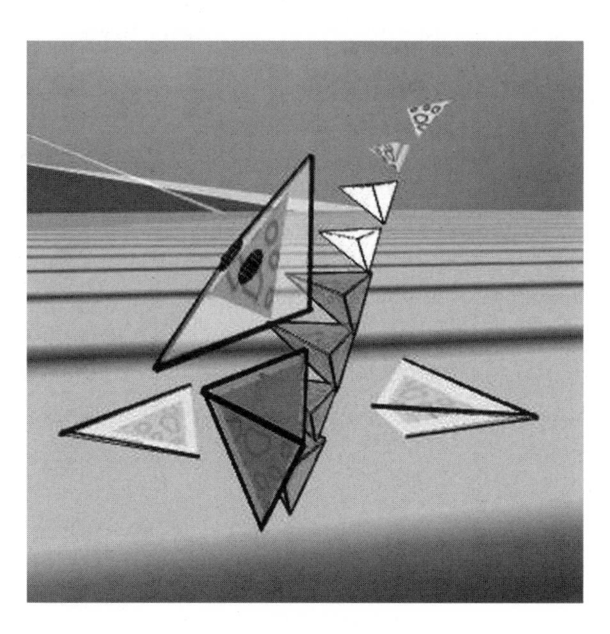

| **Figure 13.2** |

The Thing. Courtesy Dave Pape.

character. Human beings personify and react emotionally to their cars and computers; I therefore assumed that they would be equally willing to react to a computer creature that itself appeared emotional and directly solicited an emotional response.

The Thing Growing depends to a certain extent on each individual user, and so the following is a generalized account of the project. In this account and in the discussion that follows I include some users' comments on their experiences. These comments were gathered from videotaped documentation of a show in March 2000 at the EVL. The users refer to the Thing as she or it and to the piece as a film or a game.

The Thing Growing starts with the user standing in the CAVE facing a large (virtual) archway with a black void beyond. As she drives through the archway, she is teleported onto an immense plain. A voice-over prompts her to go to a shed in the distance. Inside the shed is a box. A key appears. If the user clicks on it the box bursts open, blowing the shed away in the process, and the Thing leaps out. The Thing is as big as the user and looks a little like a dragon. It dances around and shouts, "I'm free! You freed me!" Then it bows before the user saying, "I love you!" In this section, the two main protagonists, the user and the Thing, are introduced, and the Thing declares its interest in the user. Because the user frees the Thing and is warmly thanked and loved, the ideal user feels a sense of satisfaction at doing good and is drawn to her amusing and devoted new friend.

Next the Thing tells the user that it is going to teach her a special dance, a dance for the two of them. The Thing demonstrates a dance step and asks the user to copy it by moving her own physical body (figure 13.3). The trackers on the user's body tell the program (and the Thing) how the user is doing. The Thing praises or criticizes the dance. If the user gets fed up and navigates away the Thing runs after her and relentlessly coaxes, whines, or threatens her into continuing to dance. In this section the Thing progressively reveals that it is dominating and controlling. At first it praises the user's dancing; later it begins to nitpick and complain that the user isn't really trying. The user feels increasingly invaded by the Thing, which is always a little too close for comfort, and grows increasingly sick of it. When it finally flies into a temper and runs off, the user is relieved. Here are some users' comments about the dancing section:

I loved the attitude of the character.

The character was very manipulative.

I think the illusion of intelligence was interesting.

| **Figure 13.3** |

User dancing with the Thing. Courtesy Natt Mintrasak.

I really enjoyed the dancing section. When I got it right she said, "Good job," and when I didn't she was so mean.

The user's relief is short-lived. Once the Thing has gone, rocks on the plain come alive and herd and stalk the user. One of them rears up and traps her. Seconds later the Thing arrives to tell the user that it will get her out from under the rock if she is nice to it. The Thing teases the user, trying to force her to dance in exchange for freedom by telling her that the rock is dropping acid on her head. If the user dances she is freed, if she refuses the Thing eventually relents and lets her go. Once the user is released, the Thing suggests that now it should copy as the user dances. As the user moves her head and arms, the *Thing* follows suit. Almost universally users enjoy this moment when the virtual character seems to come under their control.

This moment of togetherness is swiftly broken, however, by thunder, lightning and a terrifying God-like voice demanding, "What's going on here?" Another voice from the sky screams, "We must stop this evil!" The plain cracks. The user and the Thing plunge through the earth into an eerie red underground environment. They are welcomed by the Thing's four cousins. Initially the cousins seem friendly, but the Thing is frightened and with good cause. The cousins trap the user and berate the Thing for entering into a relationship with a "meat object" (i.e., the user). They beat the Thing severely and rip off one of its arms. Then they exit mouthing dark threats. The Thing reveals that it is heresy for one of its kind to dance with a meat object. The Thing produces a gun, and it becomes the user's job to blast them out of prison and then to kill the cousins. The user is usually only too willing to run about and shoot at the cousins. They are evidently "baddies," besides which it is a moment of agency for the user in a story which, up to now, has consisted of being trapped and bullied. My goal in this section is to put the Thing and the user on the same side against a common enemy.

Finally all the cousins are killed or have escaped. The Thing and the user are alone again. But now the user has a gun. The entire piece is designed for this moment. The Thing suddenly realizes that the user could turn the gun on it. Depending on the user's actions up to this point, the Thing may challenge the user to kill it or beg for its life. The question for the user is, should she kill the Thing or not? There are two endings, one for each alternative. Neither, however, allows the user ultimately to escape the trap of this clinging relationship:

And I also wondered, should I shoot her. She was really challenging me, saying, "Oh you're not going to do it."

The humor of the guy inside the computer was unbelievable. When I shot her, when I shot her she was so funny.

The whole experience lasts about fifteen minutes. Although the experience is organized around a fairly linear narrative, the Thing responds differently to the user depending on her actions. The extent to which the user is willing to enter the experience also affects the experience:

I was talking back to it. It said, "You're doing a bad job." And I said, "I'm trying." I became very interactive, which I think you should do, because if you don't you won't enjoy the film.

The Thing Growing was a collaboration between myself and Dave Pape. I created the basic story; he created the basic code. My thinking about the visual style was influenced

———

by his earlier VR work, which stressed simplicity and interactivity. He was my mentor as I learned to write code and my advisor as I designed code for this specific project. Together, through an iterative testing process, we worked out the user interface for the project.

The Place of Visual Design in the Authoring Process

The visual decisions I made in creating *The Thing Growing* were an integral part of the authoring process. The first aesthetic issue I will discuss revolves around designing for a real-time medium. Then I will contrast designing for film and designing for VR.

Low Resolution versus Real-Time Interaction: A Trade-Off

There is a trade-off between the interactive, real-time aspect of virtual reality applications and the resolution of the graphics. Although the visual experience of being immersed in 3-D graphics stuns most new users of the CAVE, people are used to video and film, which have much higher resolution, and have a hard time understanding why the CAVE cannot similarly reproduce "real life" but in 3-D (like the Holodeck and VR as seen on television). The technology does not currently exist (except in a very experimental sense) to capture still or moving 3-D images in the way we capture film or video and turn them into 3-D objects in the CAVE (Czernuszenko et al., 1999). Therefore virtual worlds are typically created with computer graphics models and programming. The graphics in the CAVE are drawn from the user's point of view, changing as she turns her head or travels through the virtual world. The user can also interact with objects in the virtual world using the buttons on the wand. An immediate response to the user can be accomplished only by rendering the graphics in real time. This means that the computer has to update the 3-D graphics at twice the frame rate as in film or video (it draws a frame for each eye); otherwise we see jerky motion like a slowed-down film. There is a limit to how many pixels even the most powerful computer can draw in one-fiftieth of a second. To compare, one frame of 3-D computer animation in a film like *Toy Story* is of a much higher resolution than that used in VR applications but can take the same powerful computer hours to render (draw).

Nevertheless, many CAVE applications strive to be as realistic and highly resolved as possible. Much of the VR artwork I saw when I arrived at EVL focused on the visual magic that a 3-D virtual environment can accomplish. People created fabulous worlds, but when they showed them, the creators would stand in the CAVE, explain the world, navigate through it, and demonstrate any interaction involved (which was usually minimal and added as an afterthought). I wanted to create user-centered fiction, and for me the essence of the experience was the interaction. I wanted the user to have direct contact with my virtual experience and for my virtual experience to have direct contact with her. I wanted the world to explain itself and the way of interacting with it to the user. But it seemed that visual complexity had a negative impact on interactive complexity.

I realized that at some level this was a time issue. It is time consuming to construct near-photo-realistic models. It is time consuming to construct and animate elaborate 3-D characters (Joiner 1998, 156). It is equally time consuming to construct complex interaction, smart environments, and characters with intelligent behavior. It seems logical to construct the set and characters before moving on to the behavior and interaction, but my fear was that too much attention to the details of the former meant that the latter would always get shortchanged. I determined to prioritize the interactive aspects of the work over the visuals, which would be stripped down and simple, designed solely to further my plot and to facilitate my interaction.

I have also become convinced that high visual resolution of landscapes, buildings, and characters is only one axis of the sense of reality that a virtual environment can provide. When I speak of reality I mean the extent to which the user feels *present* and *immersed* in the virtual environment, two qualities that have been studied by Mel Slater and S. Wilbur (1997, 603–616). Writing about Disney's VR project *Aladdin*, Pausch et al. (1996) say, "We suspect that the limited believability of our first system's characters is due to low fidelity"(197). In contrast my experience is that simply rendered characters and environments that respond to the user intelligently are very believable. In *The Thing Growing* there are no elaborately modeled 3-D structures and few texture maps. The choice here is to make the objects in the space read symbolically rather than literally and is analogous to a simple set design in a theater. In this case the "realism" of the play lies in the acting and in the psychological impact, not in the literal re-creation of a Regency drawing room or a waterfront.

Everything in *The Thing Growing* is designed around the psychological state I am trying to create in the user. To this end the initial visual impact of the project suggests illustrations for a child's storybook; this is reinforced by a voice-over that also sounds like someone reading a children's story. The goal here is to lure users into a false sense of security so that they will be inclined to like the Thing. But this visual style also contains a hint: At the heart of fairy tales there is always menace. The Thing is similarly very simple visually. Its body is composed of multicolored semitransparent pyramids that do not connect (see figure 13.4). The body parts are animated, however, with motion tracking. A lifelike movement results, creating the strong illusion of an autonomous being: The illusion is not broken by parts of the body joining badly. Here

| **Figure 13.4** |

The Thing on the plain. Courtesy Dave Pape.

realism is displaced from a visually detailed rendition of a mythical creature (the Thing is like a dragon) to a detailed rendition of movement and to the Thing's ability to interact physically with the user. As I mentioned above, the user can see her own body in the 3-D environment and therefore retains real-world norms such as a sense of her own space. The Thing is a peer of the user in terms of size, and its physical presence is used to intimidate. It constantly invades the user's space as part of its campaign of bullying intimacy, bringing into play a very different aspect of reality:

It's so real dancing there in my face, insulting me making me mad. Actually I took a couple of swings at it.

In this project the user is drawn into the process of completing the visual work. She is expected to decode the landscapes and objects: the innocence of the plain, the mystery of the box, the danger of the red underworld. She is also required to quite literally finish the Thing's body, using the naturalistic motion as a guide to complete a being with head, arms, body and tail. This adds a level of interactivity before the more literal interactive aspects of the project even begin and from the start demands the user's cooperation and complicity in the fiction that is being created.

Film Design versus VR Design

The Frame

A powerful tool for filmmakers is control of the frame. Not only does the filmmaker control exactly what the viewer sees at any point, but camera movements have evolved a second order of meaning. For example: "The camera's moving in, either in a single take or between cuts, signals to the audience that the emotional charge of the scene is rising" (Delaney 1994, 85). Film, video, and television have a constantly evolving set of visual conventions that convey meaning: wavy lines to mean a dream sequence, jerky-point-of-view shots to convey immediacy and fear, and so on.

In VR the frame does not exist. The designer does not have control over the camera, since, in effect, the camera is the user, who can navigate through the virtual landscape looking where she wants, when she wants. The designer-programmer does, however, have control over the navigation, and this becomes an important part of the visual planning. Navigation code takes information from the joystick and uses it to compute where the user is moving in the virtual world and therefore what should be drawn. The programmer, however, determines how to translate that joystick information into movement. I can decide whether to allow the user to travel at 100 feet per second or 1 foot per second. I can impose a direction and speed on the user or decide not to allow her to navigate at all, whatever she does with the joystick (in this case she can still physically walk within the space of the CAVE). Controlling aspects of the navigation programmatically is used both to make the environment more user friendly and in accordance with the exigencies of the storyline (Pausch et al. 1996, 200).

In *The Thing Growing* I manipulated the user's ability to navigate and the speed of that navigation at many points of the narrative. At the beginning there is an audio hint that the user should go to the shed. Once the user gets close to the shed door, however, the free navigation is disabled, and the user is dragged into the shed on a fixed path. This is partly to help novice navigators, who find it hard to get through the relatively small doorway, and partly to force the user into the next scene. The virtual walls of the shed align with the screens of the CAVE and, while the user remains inside the shed, the navigation is kept off; this is because in a confined space it is easy for the novice user to spend time being bounced off walls by the collision detection. Instead I place the user exactly in front of the jumping box, so she knows on what her attention should be focused. The user's ability to navigate is also turned off when she is trapped first by the rock and again by the four cousins. In each case where the navigation is off, I tried to make a visible reason for this disabling. At times, however, when I wanted the user to listen to the Thing, I also slowed the navigation to a crawl. Since my story is about being confined and hemmed into a claustrophobic relationship, it made sense to me to manipulate the navigation. I hoped that the sense of not being able to move, of having power removed at some points, would heighten the user's frustration and feelings of not being in control:

I wanted to take control. And the game wouldn't let me take control, like most games do.

Although I cannot control where the user is going when she is navigating freely, information from the tracking system does tell me where the user is. Therefore, I can bring objects to the user. For example, the Thing is moved relative to the user most of the time rather than having an absolute position of its own in the coordinate system. It takes up a position in front of the user, at her side, or behind her; only when it goes off in a huff does it break out of this relative system and pick a place in the environment to which to go. Likewise I need a crack to open under the user's feet to precipitate her from the plain where the dancing takes place to the underworld of the four cousins. I don't have to maneuver the user into the place that this crack will be, I simply put the crack where the user is. I can't make the user fall through the real floor into the underworld, but I can make the world move up past the user to give the illusion of falling. Thus part of the set I design is not fixed in a coordinate system but positioned on the fly relative to the user. Managing scenic elements that have a fixed place in the coordinate system and those that will be placed relative to the user's current position becomes a new design issue.

Timing

Camera movement and editing are the tools a film or video maker uses to control the timing of a piece. In film, and also in any linear narrative, the author has absolute control over the timing of events, and this is considered a crucial part of building and releasing narrative tension. At the SIGGRAPH panel Fiction 2000, Andrew Glassner and C. Wong (1999) suggested that, for the construction of interesting computer fiction, a contract has to be forged between the author and user. One point in that contract is that the author will control the sequencing of events and the creation of a causal chain of action.

A 3-D virtual environment, however, is more usually seen in terms of space; an interactive event will occur based on the user's proximity to something in the scene rather than on timing. Thus in *The Thing Growing*, the user has to move into the shed before she can release the Thing from the box. As I approached this piece, however, I did not want to lose all the advantages of pacing and surprise that control over timing gives an author. So in the scene with the user in the shed, there is a timed element. The user's entering the shed triggers the box to jump up and down. A voice comes out of the box begging to be released. A problem is posed for the user, but the solution is not given. The problem hangs in the air for a few seconds, while frustration builds. Then a key appears and announces that it can open the box. The user has to click on the key to activate it. More sophisticated users click on it before it has had time to announce its function, indicating their frustration level. Those less used to VR and computers, may not click on the key automatically. Therefore the key prompts the user to click on it, which will make it open the box, precipitating the next stage in the story.

My general solution to the timing issue was to build the story in stages. The story moves from one stage to another either because the user has made certain choices, or because the time allotted to a particular stage is up. This keeps the story moving forward, but within each stage there is room for choice and interaction. For example, during the dancing stage, the program counts the number of times the user refuses to move or constantly runs away from the Thing, storing the number of disobediances. As the number rises, the Thing's mood worsens. If it gets too high, it triggers the Thing to go off in a huff. In case the user does not reach this number of disobediences, however, I also established a fixed maximum duration for the dance stage after which time the Thing also takes off in a huff. The sequence in which the user has the gun and can shoot the cousins is likewise terminated either by the user's killing them all or after a fixed interval. This method of creating discreet stages or scenes was also used by the creators of *Kidsroom*, a smart room for children that involved them in an interactive story. One of their findings is that it is important not to create interactive cul-de-sacs, where the story will continue only if the user performs a specific action (Bobick et al. 1990, 390). Instead the program should check to see if the user performs the action within a certain time frame, and if not, should autonomously move the story on.

Cuts and Transitions

In film and video, conventions have also been established that mean experienced viewers do not get lost in the face of wild transitions between scenes or camera angles. In VR it is trivial to move the user from one virtual scene to another, the equivalent of cutting from scene to scene in a film. It can also be completely disorienting, however, since there are no established VR conventions for understanding what has happened. Working on this piece I came to believe that in immersive VR, the scene transitions have to be foreshadowed in order to work: The user has to be aware that a change is going to happen and have an idea why it's going to happen.

A simple example of a marked transition occurs at the very beginning of *The Thing Growing* when the user is instructed to follow arrows that appear on the floor in front of her and point through an archway into complete blackness. The threshold created by the archway indicates that she will be entering something, although she doesn't know what. A more explicitly foreshadowed transition occurs between the scene on the plain, where the user dances with the Thing, and the underworld, where the four cousins threaten them. During the development process, we had a very simple cut at this point: At one moment the user was on the plain with the Thing, at the next in a different place with the four cousins approaching. Users did not have time to figure out what had happened, and this affected their understanding of the story. Some thought that the cousins were coming to rescue them. I therefore added signs that presaged the change. There is lightning and a voice roaring in the sky that reveals someone is not pleased with what is happening (the relationship between the Thing and the user). The plain cracking and tunnel/pit appearing under the user and three or four seconds of apparent falling into the new environment give the user time to appreciate that she and the Thing have fallen into the world and hands of some other Things who are hostile (figure 13.5).

| **Figure 13.5** |

An avatar of the user and the Thing face the four cousins.

I believe immersive VR will develop an evolving lexicon of visual norms, metaphors, and conventions that will include attention to navigation, transitions, triggering events by proximity *and* timed interval, and building virtual sets that combine elements in the fixed coordinate system and elements constructed on the fly in relation to the user. As computers become more powerful, the issue of visual resolution will disappear. I hope, however, that the possibility of photo-realistic 3-D effects will not lead to the demise of all other visual styles possible in the VR environment.

Coding versus Writing

At one level of organization, writing *The Thing Growing* was exactly like writing a short story or film: I had to develop a plot and delineate a character through dialog. In both writing and programming I am organizing what happens, when and how, to tell my story most effectively. The computer code that implements the story can be seen as a combination of the following aspects of a play: a description of all the costumes, sets, and props and how they should be built; the book that contains all the lighting cues, set and prop changes; all the director's instructions to the actors for their entrances and exits and stage business; and all the lines that the actors will actually speak. In addition, the code needs to define any physics the creator may want in her world (gravity, solid walls, etc.), supply all the intelligence she may want her virtual actors to display, and make provisions for the choices of an unknown and continually changing human participant in the play: the user.

XP: VR Authoring System

The Thing Growing was written in *XP*, a VR authoring toolkit designed by Pape and his colleagues at EVL to facilitate the construction of art applications in the CAVE (Pape et al. 1998). The toolkit handles a number of activities common to VR environments, such as assembling 3-D models into a world, collision detection, navigation, triggering and detecting events, and passing messages in response to events. A useful tool in *XP* is a text file in which the creator writes down all the objects that exist in the virtual world being created and instructions for them. For example, the creator writes down all the pieces of the set, their positions in the coordinate system, and any movement they might make. The narrative of *The Thing Growing* was created using such text files: Timed sequences were intercut with the interactive episodes; the narrative flow as a whole was structured using triggers based on time, user proximity, or the completion of specific events; all the Thing's dialog and instructions for movement were listed.

The text file served as production manager for the story, which could therefore be easily edited and changed. At this level, the process was very writerly.

Underlying the text file, however, is the code that parses the text file into a set of instructions for creating and manipulating graphics and audio. This set of instructions is taken by the compiler and turned into machine code, until the computer arrives at the ones and zeros that the hardware understands and applies to pixels on a screen and to audio drivers. The further down this chain one goes, the less writerly the process. To create *The Thing Growing*, I had to write the code as well as manipulate text files. The *XP* system creates generic tools that can be used to build a generic world. For the specifics of my story I had to extend the system, and with Pape's help, I built intelligence for the Thing and for other elements of this virtual environment.

Writing with Mathematics

The first way that coding is not like writing is that one must use mathematical calculations to make things move. I said in the last section that the instructions for moving the Thing are listed in the text file. These instructions say things like, "stay close to the user," "get in the user's face," "run and hide." The underlying code takes these verbal instructions and turns them into the actual movement, which is calculated by, for example, figuring out where the user is and placing the Thing at a coordinate position close to that. How the Thing moves is an essential part of its character. The Thing is therefore not only the result of a writerly operation. It could be argued that the intent is writerly and that it is merely a technical detail to make the Thing move in accordance with the intent. But as any writer knows the real work comes when one moves from the intent to the writing that actualizes the intent. Here that process is profoundly mathematical.

Iteration

A second way that coding is not like writing is that it is iterative. For every frame, the code checks to see where the user has moved or where she is looking in order to update the graphics. But it also performs other checks and operations that create activity or respond to the user. So it's as if on every frame the code scans the story and all the alternatives built into the story to give the user the appropriate next frame in terms of both visuals and narrative. I can illustrate this best by explaining how the Thing is constructed as a responsive agent with both a body and a brain.

The Thing's body has three basic elements: a 3-D computer graphics model of the body parts, animation for those parts, and a voice. As I mentioned above, the Thing's body is made up of semitransparent pyramids. We used motion capture to animate the body parts naturalistically, but we also had to create the algorithms that moved the body as a whole about in the environment. The Thing's voice is prerecorded. Based on the storyboard (the plot as designed, prior to programming), we recorded hundreds of pieces of dialog for its voice. Sometimes its speeches are scripted and do not vary (for example, when it is freed from the box in which it is trapped). But mostly it speaks in response to the user (for example, when it is teaching the user to dance). For the dance section we recorded different versions of words of praise, encouragement, criticism, explanation. We also recorded the different types of utterance in different moods: happy, manic, sad, and angry. Each phrase lasts a few seconds. We built up a library of *actions* for the Thing that consist of short, motion-captured sequences linked to corresponding dialog. Each action lasts a few seconds and also contains information on how the body as a whole moves.

When the piece is running, the program sweeps through all its instructions about thirty times a second to see which ones to implement. Therefore thirty times a second, the Thing's brain checks to see what it should tell the Thing's body to do. First it checks to see if it is in the middle of an action, and if it is, it lets the body finish the action. Actions, however, can be from one to five seconds long. Sometimes that is too long a response time, so there is also a mechanism for interrupting the current action. Whether the Thing is willing to be interrupted depends on the storyline. If the brain determines that the body has reached the end of an action—it has said its piece of dialog and done its movement or should be interrupted—the brain has to select the next appropriate action. It makes the selection according to the point in the narrative, the user's actions, and the Thing's own emotional state, all of which it checks by looking at information gathered during the last action. It is the task of the body to interpolate smoothly between the end of one action and the beginning of the next—a task made much simpler because of the simple graphics. Figure 13.6 presents a greatly simplified schematic of the Thing's decision-making process.

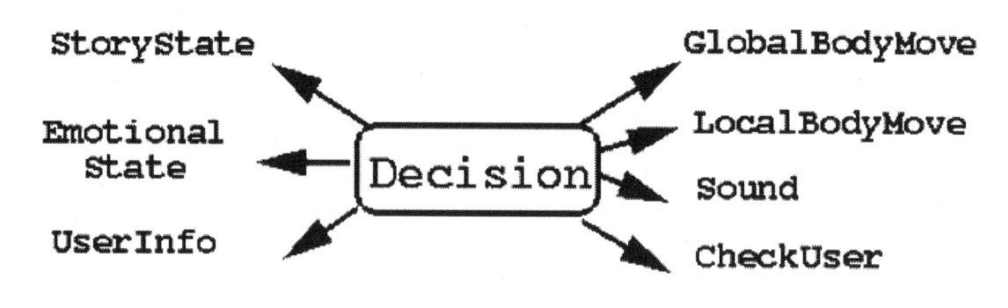

| Figure 13.6 |

Simplified diagram of the Thing's decision-making process.

Narrative Constraints

The narrative becomes a very useful tool for constraining the kind of action the brain can pick, thus limiting the number of actions we had to program. For example, when the Thing is attempting to teach the user to dance, it has a basic routine to follow. It demonstrates each part of the dance, then observes or joins the user as she copies the movement. As the user does so, information on whether the user is dancing correctly is recorded so that it can be accessed by the brain's checking system. The Thing may admonish, encourage, or praise the user according to her behavior and its own mood. It may repeat a part of the dance that the user is doing incorrectly, or it may teach another step. This routine is interrupted if the user tries to run away, and behavior is triggered to make the Thing run after the user and plead with or scold her to continue the dance. Each type of response—"user_danced_well," "user_ran_away," "new_dance_step"—corresponds to a store filled with possible actions. The brain can pull an action out of the store sequentially—for scripted moments in the story—or randomly, or by mood. One of the major headaches of writing for the Thing, however, was to ensure that anything it said fit in with the last thing it said and the next thing it was going to say. This, in essence, meant keeping track of the hundreds of different phrases and the millions of ways they could be combined. The constraints of the narrative and basic routines like the one described above made this possible.

Scripting the User

Another important difference in writing for VR is that the code that drives VR must contain all the alternatives that may be needed depending on what the user does; in effect it has to create an empty protagonistic space for *anyuser* to act in and be ready to respond to *anyuser*'s actions. In many ways it is fortuitous that our virtual character is dominating, because a very proactive character can both tell and control the story—the user blames her lack of control on this character and on her own inability to wrest control from it, not on the limitations of the program. And of course the program is limited. We cannot write code for every possible action a user may take. We need to find ways to constrain those actions into a manageable subset that the program can react to. Janet Murray (1997) refers to this process as "scripting the interactor" (79). Pinning down the human subject and her responses to the Thing was not a one-shot deal. The creation of a successful empty protagonistic space depended on testing that space with a variety of users and watching what they did in it. Then we adjusted, refined, and added to the Thing's functionality so that it had responses that could fold the users' different reactions back into our narrative thread. As writers need editors to help them refine their work, programmers need users to test their applications.

The iterative process of user-testing and changing the code also honed the story, since watching users made it very clear when people were not really "getting" the narrative message we wanted to send. The culmination of the story is to bring the user to a point where she can destroy the Thing if she chooses. A major problem was creating enough ambivalence at this moment. In early versions virtually everyone shot the Thing. We therefore adjusted the program to try to make the Thing more likeable. We added a section in which the user moves and the Thing mimics her. Many users liked this moment of control, when they are mirrored by the other, when the other is doing what they do for once. Adding this forty-second section took at least three weeks of coding, and again the process was intensely mathematical. We also clarified the story of the forbidden nature of the love affair. We made the cousins meaner and the Thing more pitiful as it begs the user to protect it from them. If the user doesn't kill the cousins, the Thing becomes hysterical, because she is not protecting it. If the user does kill the cousins, the Thing is aghast at the slaughter of its family and the bloodthirsty side of the user the killings reveal. In each case we provided reactions to the user that we hope channel her to the moment of truth: killing or not killing her "friend" (this is how many users referred to the Thing), with the maximum amount of ambivalence about her own actions. After all this, fewer users killed the Thing.

Creating both the personality and intelligence of a virtual creature that will draw a user through a semiscripted plot is a new type of challenge for a writer compared to those she faces in creating a traditional, linear, noninteractive story. Having a continuous loop of logical operations rather than a linear narrative as the driver of one's authoring process is also very different. Instead of writing a paragraph to illustrate a specific point, one is writing a function that will perform an operation, and one wants that function to be as generic at possible so that one can reuse it (as reuse of the function saves programming time). Instead of assembling the paragraphs into a whole, one is calling functions depending on if-statements. Very often one is using mathematics to describe actions or plot devices. Code and interface testing are an integral part of these processes. Many artists who work in VR are not programmers but work with programmers. At this point in VR's history, I believe it is important for artists and writers to be programmers for vital immersive VR dramas to emerge. The tool of programming is too qualitatively different from other visual and literary tools for the real power of this medium to be uncovered without the creator's understanding it.

Control Issues

CAVE systems are very costly and have a very large architectural footprint. Worldwide there are only a handful of CAVEs open to the public in art institutions or museums (Roussos 1999, 34). EVL has developed a small, drafting-table version of the CAVE, the ImmersaDesk, which is portable and can be taken to shows (Czernuszenko et al., 1997, 46–49). Researchers are also experimenting with lower-cost VR projection systems using PCs. Currently, however, few people have experienced immersive VR, and so they come to the experience as novices. It is very typical when CAVE VR is shown that an expert navigator drives a group of users through the VR experience. The whole point of *The Thing Growing*, however, is for the user to be the protagonist; therefore the user *must* have control over the wand and navigate through the experience on her own. Therefore I created an introductory scene in which she learns the basics of navigation and interaction. A human helper sets the user up with the glasses, wands, and tracking sensors and then steps away. The user finds herself facing a billboard on which is a diagram of the wand showing the joystick and three buttons. Text instructs the user to push the left button to begin and then explains how to use the joystick to drive. In this introduction/instruction scene the user practices navigating by following arrows that lead her to the archway. The user therefore drives herself through the archway and into the experience. Later in the piece, the Thing itself gives the user explicit instructions on what to do.

Although the user is in control of the hardware, there are also decisions to be made about how far she should be in control of the VR experience. Kelso, Weybrauch, and Bates (1993) imagine the beginning of an interactive fiction:

> You find yourself immersed in a fantasy world with exciting characters and the possibility of many adventures. Although you control your own direction by choosing each action you take, you are confident that your experience will be good, because a master interactive story-teller subtly controls your destiny. (1)

A fantasy world with the possibility of many adventures assumes a narrative that will branch in many directions, with a proliferation of plots, characters, and scenes. Although this may one day be possible, when we were planning *The Thing Growing* we dismissed this type of branching narrative for two reasons. First, we considered it too time consuming and expensive to create a multiplicity of complex scenes in VR. Joiner (1998) also makes this point in relation to the construction of "high-production-value graphical adventures" (156). Second, we wanted to take advantage of traditional narrative devices, which garner much of their power from the control of pacing, surprise, and the construction of a rising curve of interest. Our thinking received support at the SIGGRAPH 99 panel Fiction 2000, at which Glassner blasted the branching-narrative strategy, suggesting that too often it gives the user shallow choices about which she is unlikely to care. *The Thing Growing* has a much simpler goal. Instead of creating multiple stories, we create the story of one relationship and focus all the program's complexity into the virtual character. The computer-controlled character, the Thing, is programmed with a multiplicity of reactions to play opposite the user. Here, we in effect, reintroduce the concept of branching, but instead of having branching narrative trails, we put all that complexity into one character who has branching behaviors. The user's sense of interaction and agency comes from her dealings with a character who responds to her meaningfully.

Brenda Laurel also argues against the idea that a VR interaction fiction is intrinsically more interesting the more control the user has. In *Computers as Theater* (1993) she talks about the art of improvisation, explaining that creating interesting, dramatic experiences on the fly is not something that most people are trained to do, therefore "[a] system in which people are encouraged to do whatever they want will probably not produce pleasant experiences" (101). Laurel introduces the idea of constraints that can be built invisibly into the activity or experience. They control the experience but are so embedded in it that the user accepts, benefits from, and uses the control. "If the escape key is defined as a self-destruct mechanism, for instance, the constraint against pressing it in the course of flying a mimetic spaceship is intrinsic to the action" (103).

Our narrative served as such a constraint. The project as a whole has a storyline that follows an arrow of time and has the very traditional bridge structure of plays and films: Act 1 introduces the protagonists and the goal; Act 2 revolves around struggles to reach the goal; Act 3 resolves those struggles. The major difference between this case and plays or films is that in each act, the protagonist is engaged in interaction. The narrative as a whole is moved on either as a result of the user's choices or by time. Because the story and the Thing have a certain amount of autonomy and a goal, the user is not the only active agent in the narrative. Therefore she can be surprised, which in fiction is a very good thing. At some times in the story she may be powerless to act, but at other times she is urged to act; the piece as a whole is designed to maximize her sense of agency and immersion in the narrative and to give her a degree of free will.

Conclusion

The final version of *The Thing Growing* was finished in March 2000, and a Japanese translation was exhibited in Tokyo in June 2001, but since 1998 we have shown the project (at first as a work in progress) in a variety of venues, with audiences that varied from very sophisticated to relatively naive with respect to VR technology and environments. Confirmation that the users become engaged in the story and with the Thing comes as they talk back to it. Toward the end of her experience one user said, "I want out of this relationship." We have also been told that the Thing has presence; that it is oddly feminine; that it is like a whiny child; that it is like people we know; and that the relationship is like a marriage.

I have derived the following lessons and guidelines for creating immersive VR fiction from my experience in working on this project. A realistic virtual character that seems imbued with life can be created with relatively simple graphics if its level of responsiveness to the user and apparent intelligence is high. In more general terms *The Thing Growing* project suggests that very high graphics resolution and a complex 3-D visual environment is only one factor that may make a user feel present and immersed in a virtual environment. Established media have established conventions that the audience understands; these conventions are manipulated with great subtlety over time. VR is an experimental medium with no such conventions. Since users tend to be unfamiliar with and a little wary of it, I think both the form and the content need to be very clear and self-explanatory, to verge even on the obvious. Transitions between scenes have to be handled with particular clarity so that users do not become disoriented. As immersive VR and immersive VR fictions develop, I believe the following are the formal areas in which we will see the evolution of second-order meanings: navigation techniques, transition techniques, event-triggering techniques, and the design of worlds with both fixed objects and those that are positioned on the fly with relation to the user's position. I also hope to see the development of lively debates about the level of user control in VR and the impact of decisions about user control on specific applications. Finally, it is my firm belief that we need creative teams that include not only artists, writers, programmers, and artificial intelligence programmers but also people with multiple skill sets in these areas, to push this new medium toward its creative potential.

Acknowledgments

The Thing Growing was developed at the Electronic Visualization Laboratory at the University of Illinois at Chicago. CAVE and ImmersaDesk are trademarks of the Board of Trustees of the University of Illinois.

Works Cited

Bobick, A., S. Intille, J. Davis, F. Baird, C. Pinhanez, L. Campbell, Y. Ivanov, A. Schütte, and A. Wilson. 1994. "The KidsRoom: A Perceptually-Based Interactive and Immersive Story Environment." *Presence* 8:367–391.

Cruz-Neira, C., D. J. Sandin, T. A. DeFanti, R. V. Kenyon, and J. C. Hart. 1992. "The CAVE: Audio Visual Experience Automatic Virtual Environment." *Communications of the ACM* 35, no. 6:65–72.

Czernuszenko, M., D. Pape, D. Sandin, T. DeFanti, G. L. Dawe, and M. D. Brown. 1997. "The ImmersaDesk and Infinity Wall Projection-Based Virtual Reality Displays." *Computer Graphics* 31, no. 2:46–49.

Czernuszenko, M., D. Sandin, A. Johnson, and T. DeFanti. 1999. "Modeling 3D Scenes from Video." *The Visual Computer Journal* 15, nos. 7–8:341–348.

Delaney, S. 1994. *Silent Interviews.* Middletown, CT: Weslyan University Press.

Glassner, A., and C. Wong. 1999. "Fiction 2000: Technology, Tradition, and the Essence of Story." In *SIGGRAPH '99 Conference Abstracts and Applications, Los Angeles, CA, August 1999*, #161. New York: ACM SIGGRAPH.

Joiner, David Talin. 1998. "Real Interactivity in Interactive Entertainment." In *Digital Illusion*, ed. Clark Dodsworth Jr., 151–159. New York: ACM Press.

Kelso, M., P. Weybrauch, and J. Bates. 1993. "Dramatic Presence." *Presence* 2:1–15.

Laurel, B. 1993. *Computers as Theater.* Menlo Park, CA: Addison-Wesley.

Mateas, M. 2001. "A Preliminary Poetics for Interactive Drama and Games." In *SIGGRAPH 2001 Electronic Art and Animation Catalog, Los Angeles CA, August 2001*, 51–58. New York: ACM SIGGRAPH.

Murray, J. 1997. *Hamlet on the Holodeck: The Future of Narrative in CyberSpace.* New York: Simon & Schuster.

Pape, D., T. Imai, J. Anstey, M. Roussou, and T. DeFanti. "XP: An Authoring System for Immersive Art Exhibitions." In *Proceedings of Fourth International Conference on Virtual Systems and Multimedia, Gifu, Japan, November 18–20, 1998*, ed. Hal Thwaites, 528–533. Burke, VA: IOS Press.

Pausch, R., J. Snoddy, R. Taylor, S. Watson, and E. Haseltine 1996. "Disney's Aladdin: First Steps toward Storytelling in Virtual Reality." In *Proceedings of SIGGRAPH 96, New Orleans, LA, August 4–9, 1996*, 193–203. New York: ACM SIGGRAPH.

Roussos, M. 1999. "Incorporating Immersive Projection-Based Virtual Reality in Public Spaces." In *Proceedings of the Third International Immersive Projection Technology Workshop, Stuttgart, Germany, 10–11 May 1999*, ed. Hans-Joerg Bullinger, 33–40. Berlin: Springer-Verlag.

Slater, M., and S. Wilbur. 1997. "A Framework for Immersive Virtual Environments (FIVE): Speculations on the Role of Presence in Virtual Environments." *Presence* 6:603–616.

Contributors

Josephine Anstey is a virtual reality artist, a writer, and video maker. Her virtual reality pieces, collaborations with Dave Pape, have been shown in museums and festivals in the United States, Europe, and Japan. In 1996 her short story "Dwayne Loves Johnny Coggio" won the Chelsea Award for short fiction. Since 1983 she has collaborated with Julie Zando on a series of videos, many of which are in the permanent collection of the Museum of Modern Art.

Jan Baetens is assistant professor of visual culture at the Institute for Cultural Studies of the University of Leuven (Belgium) and in the department of Visual Studies at the University of Maastricht (Holland). He is the editor of the on-line journal *Image (&) Narrative* <www.imageandnarrative.be> and has written or edited some thirty volumes, most of them in French. In English, he edited, with José Lambert, *The Future of Cultural Studies* (Louvain University Press, 2000). He is also a poet. His book "Self-Service" (copublished by Casa Fernando Pessoa and the Éditions Fréon) is available in a trilingual (French, English, Portuguese) version.

Nancy Barta-Smith is associate professor of English at Slippery Rock University. Her research interests include the rhetoric of sciences, legal rhetoric, technical and scientific writing, embodiment, and phenomenological approaches to literature and composition. She was the coconvenor of Slippery Rock University's Teaching, Learning, Technology Roundtable from 1998 through 2001.

Jay David Bolter is codirector of the New Media Center and Wesley Professor of New Media in the School of Literature, Communications, and Culture at the Georgia Institute of Technology. His research focuses on the computer as a new medium for verbal and visual communication. His work with computers led in 1984 to the publication of *Turing's Man: Western Culture in the Computer Age*, which was widely reviewed and translated into several foreign languages. Bolter's second book, *Writing Space: The*

Computer, Hypertext, and the History of Writing, published in 1991, examines the computer as a new medium for symbolic communication. With Michael Joyce, Bolter is the author of *Storyspace*, a program for creating hypertexts for individual use and World Wide Web publication. In his most recent book, *Remediation: Understanding New Media*, written in collaboration with Richard Grusin, Bolter explores the ways in which new digital media, such as the World Wide Web and virtual reality, borrow from and seek to rival such earlier media as television, film, photography, and print.

Helen Burgess is assistant professor in Electronic Media and Culture at Washington State University, Vancouver. She is coauthor with Robert Markley, Harrison Higgs, and Michelle Kendrick of *Red Planet: Scientific and Cultural Encounters with Mars* (University of Pennsylvania Press). Her other publications include essays on virtual reality and feminism, X-rays, and crash test dummies. She is currently working on a multimedia history of the interstate superhighway.

Alice Crawford is a doctoral candidate in communication at the University of Pittsburgh. Her dissertation investigates relations between spatial metaphors embedded in information technologies and the rhetoric around them and the manner in which urban spaces are experienced and used. She worked in information design before her current incarnation as an academic.

Danette DiMarco is associate professor of English at Slippery Rock University. Her research interests include the coordination of verbal and visual literacy, liberatory approaches to pedagogy, especially in a technological age, and postcolonial and feminist approaches to literature. She was honored with Slippery Rock University's President's Award For Excellence in Teaching in 1999. She has been a member of the university's Teaching Learning Technology Roundtable since 1998.

Jeanne Hamming, in addition to working on the ongoing *Mariner10* series, is currently writing her dissertation on masculinity, media, and the environment and teaching literature at West Virginia University. Her ongoing research and publication interests include gender studies and popular culture.

Gail E. Hawisher is professor of English and founding director of the Center for Writing Studies at the University of Illinois, Urbana-Champaign, where she teaches graduate and undergraduate courses in writing studies. Her most recent publications are the coedited *Global Literacies and the World Wide Web* and *Passions, Pedagogies, and 21st Century Technologies*, which won the Distinguished Book Award at Computers and Writing 2000. With Cynthia Selfe, she edits the international journal *Computers and Composition.*

Mary E. Hocks is assistant professor of English at Georgia State University, where she teaches digital rhetoric and writing courses and also directs the Writing across the Curriculum program. She has published articles on hypertext, feminism and technology, and multimedia writing and pedagogy. With Anne Balsamo, she serves as project director on the multimedia documentary, *Women of the World Talk Back*, originally created for the Fourth World Conference on Women in Beijing, China.

Michelle Kendrick is assistant professor of English at Washington State University, Vancouver. She teaches undergraduate and graduate courses in the Electronic Media and Culture Program. Her many articles and multimedia publications focus on new media, visual literacy, technology and subjectivity, and technologies of war. She is a director of the *Mariner10* DVD series from the University of Pennsylvania Press and is a principal coauthor of *Red Planet: Scientific and Cultural Encounters with Mars* (University of Pennsylvania Press, 2001).

Matthew G. Kirschenbaum is assistant professor of English at the University of Maryland, College Park, where he teaches digital studies and applied humanities computing. He has a particular interest in digital images and is technical editor of the electronic *William Blake Archive;* he is also the codeveloper of the Virtual Lightbox, an open-source peer-to-peer image-sharing tool. His current book is *Mechanisms.*

Kevin LaGrandeur is assistant professor of English and director of technical writing programs at New York Institute of Technology. His articles on computers and writing and on science and literature have appeared in *Computers and the Humanities*, *Computers & Texts*, *English Studies*, and *Texas Studies in Literature and Language*, as well as in other journals, online sources, and the *ERIC* database.

Carol Lipson is associate professor of writing and rhetoric at Syracuse University. With Roberta Binkley, she is coediting a book on ancient rhetorics, *Rhetoric before and beyond the Greeks*, to be published by SUNY Press. Her previous publications have focused on the history and theory of scientific and technical writing.

Robert Markley is Jackson Distinguished Chair of British Literature at West Virginia University and, in fall 2003, will become professor of English at University of Illinois, Champaign–Urbana. He is the author of five books and over fifty articles on seventeenth- and eighteenth-century literature, cultural studies, and the relations between literature and science. His following books include *Dying Planet: Mars and the Anxieties of Ecology from the Canals to Terraformation* (Duke University Press, 2004) and *Fictions of Eurocentrism: The Far East and the English Imagination 1600–1800* (Cambridge University Press, 2004). He is coauthor of *Red Planet: Scientific and Cultural*

Encounters with Mars (University of Pennsylvania Press, 2001), the first scholarly title authored for DVD-ROM, editor of *The Eighteenth Century: Theory and Interpretation*, and coeditor of *Mariner10: Educational Multimedia* for the University of Pennsylvania Press <www.mariner10.com>.

Ellen Strain is assistant professor of film and multimedia in the School of Literature, Communication, and Culture at the Georgia Institute of Technology. She teaches film history and multimedia design and is a freelance instructional technology designer. She has just finished *Public Places, Private Journeys: Ethnography, Entertainment, and the Tourist Gaze* for Rutgers University Press and is currently authoring a multimedia tool, *Project Noir*, that allows students to reedit a hypothetical, unfinished opus to learn about the effects of editing on film noir's narrational strategies. Strain also works with a division of Pearson Education on the development of instructional technologies for K–12 education.

Patricia A. Sullivan is professor of English at Purdue University, where she directs the graduate rhetoric program. Her research intersects rhetoric, technology, and feminist issues and methodology. Her two most recent books are the coedited *Electronic Literacies in the Workplace: Technologies of Writing* and the coauthored *Opening Spaces: Writing Technologies and Critical Research Practices.* She is currently at work, with Pam Takayoshi, on an edited collection titled *Labor, Writing Technologies, and the Shaping of Composition in the Academy.*

Gregory VanHoosier-Carey is R. Z. Biedenharn Associate Professor of Communication at Centenary College of Louisiana, where he teaches courses in design communication, multimedia design, cultural studies, and rhetoric. His research in new media communication explores cultural and design issues in electronic discourse and hypermedia, especially those related to computer-mediated pedagogy.

Jennifer Wiley is assistant professor of psychology at the University of Illinois at Chicago. She has conducted research on learning from text, including electronic text, and is generally concerned with the contexts and conditions that allow readers to develop understanding as they read.

Anne Frances Wysocki teaches undergraduate courses in visual communication, web development, and interactive media as well as graduate courses in technology studies and theories of new media in the Humanities Department at Michigan Technological University. Her research into visual rhetorics and the rhetorics of interactivity focuses on the relationships that writers/designers/composers construct with their audiences through various kinds of texts.

Index